One by Walking

Environment and Society

Series Editor

Douglas Vakoch

As scholars examine the environmental challenges facing humanity, they increasingly recognize that solutions require a focus on the human causes and consequences of these threats, and not merely a focus on the scientific and technical issues. To meet this need, the Environment and Society series explores a broad range of topics in environmental studies from the perspectives of the social sciences and humanities. Books in this series help the reader understand contemporary environmental concerns, while offering concrete steps to address these problems.

Books in this series include both monographs and edited volumes that are grounded in the realities of ecological issues identified by the natural sciences. Our authors and contributors come from disciplines including but not limited to anthropology, architecture, area studies, communication studies, economics, ethics, gender studies, geography, history, law, pedagogy, philosophy, political science, psychology, religious studies, sociology, and theology. To foster a constructive dialogue between these researchers and environmental scientists, the Environment and Society series publishes work that is relevant to those engaged in environmental studies, while also being of interest to scholars from the author's primary discipline.

Recent Titles in the Series

One by Walking: Transdisciplinary Mobilities and Methodologies, edited by Camilla Brudin Borg, Hanna E. Åberg, Roger Norum, Susanne Österlund-Pötzsch, and John Martin

A Regulatory Framework to Remediate the Planet: Strategies, Outcomes and Implications, by Robert Smet

Climate Diplomacy at the Crossroads: A Multidisciplinary Perspective, edited by Şuay Nİlhan Açıkalın

Political Ecologies of Futurity: Storytelling Plantation Afterlives, Climate Erasures, and Socioecological Justice, edited by Alex A. Moulton and Dylan M. Harris

Concrete Mirage: Governance, Equity, and Sustainable Cities in the United States Southwest, by Erin Heinz

Building Sustainable Agrifood Systems and Resilient Rural Communities in Japan, edited by Raymond Jussaume, Keiko Tanaka, and Kiyohiko Sakamoto

Strategic Rebuilding and Affordable Housing in the Muslim World, edited by Akel Ismail Kahera, Sultan Barakat, and Ghadanfar

Encounters with the Posthuman and the Environment, edited by İnci Bilgin Tekin and Zeynep Talay Turner

Canals in a Changing Britain: Construction, Culture, and Environment, 1760–1968, by Jules P. Gehrke

Urban Narratives About Nature: Socio-Ecological Imaginaries Between Science and Entertainment, edited by Carlos Tabernero

The Age of Capitalism, Consumer Culture, and the Collapse of Nature in the Anthropocene, by Jack Thornburg

Sustainable Energy Development: Technology and Investment, edited by Kimarie Engerman, Elizabeth A. Keenan, Dmitry Kurochkin, and Elena V. Shabliy

Making Nature Social: Towards a Relationship with Nature, by Rembrandt Zegers

Earth Polyphony: Law, Ecocriticism, and Eco-Activism in Postcolonial India, by Suhasini Vincent

Staying Together: Natureculture in a Changing World, edited by Kaushani Mondal

Learning in the Anthropocene: Reimagining Education in the Twenty-First Century, by Carl A. Maida

One by Walking

Transdisciplinary Mobilities and Methodologies

Edited by Camilla Brudin Borg, Hanna E. Åberg,
Roger Norum, Susanne Österlund-Pötzsch,
and John Martin

BLOOMSBURY ACADEMIC
NEW YORK • LONDON • OXFORD • NEW DELHI • SYDNEY

BLOOMSBURY ACADEMIC
Bloomsbury Publishing Inc, 1359 Broadway, 12th Floor, New York, NY 10018, USA
Bloomsbury Publishing Plc, 50 Bedford Square, London, WC1B 3DP, UK
Bloomsbury Publishing Ireland, 29 Earlsfort Terrace, Dublin 2, D02 AY28, Ireland

BLOOMSBURY, BLOOMSBURY ACADEMIC and the Diana logo are trademarks of Bloomsbury Publishing Plc

First published in the United States of America 2025

ISBN: HB: 979-8-21636-562-4
ePDF: 979-8-21635-198-6
eBook: 979-8-21635-197-9

Typeset by Deanta Global Publishing Services, Chennai, India

Contents

Acknowledgments

This anthology is the result of a series of workshops and meetings organized by the research network One by Walking. The network has been made possible by generous support from a number of funders: Riksbankens Jubileumsfond granted us start-up assistance for our first workshop in Vålådalen, Sweden (2022). The Joint Committee for Nordic Research Councils in the Humanities and Social Sciences (NOS-HS) awarded us funding for a series of workshops in Rosendal, Norway (2022); Korpo, Finland (2023); and Sælingsdal, Iceland (2023).

The Swedish Cultural Foundation in Finland, Letterstedtska föreningen, The Royal Gustavus Adolphus Academy for Swedish Folk Culture, and the Sven och Dagmar Saléns Foundation supported the conference and workshop "Mobile Heritage: Walking and the Environment through Future and Historical Perspectives" in Hanaholmen, Helsinki (2024), from which a number of chapters for this volume emerged.

Support for editorial and proofreading work was granted by the Department of Literature, History of Ideas, and Religion at the University of Gothenburg; the Faculty of Humanities and Profi6 Biodiverse Anthropocenes programme (RCF 336449) at the University of Oulu; and the projects "RURALEX—Knowledge in Crisis: The Dynamics of Environmental Expertise amidst Rural Change" (CHANSE and HERA / RCF 362488) and "Pohjoisen päätepisteet: Sosio-ekologisesti kestävän kehittämisen ja liikkumisennykyisyys ja tulevaisuudet arktisessa Fennoskandiassa" (RCF 339423).

We are very grateful to our editors at Lexington Books, Courtney Morales, Breasia Boyd, and their colleagues, who guided us skillfully and gently toward publishing this volume. Sophie Schlondorff was the extremely knowledgable and efficient proofreader of every word of text in this book. Thanks also go to Natural England for permission to print the "Landscape Character Wheel" and for permission to reproduce images by Andy Hughes and Jason Parsons (University of Plymouth), and for all photos contributed by the authors of this anthology themself.

Last but not least, we are grateful to the academic and artist members of the One by Walking network, our colleagues in perambulation, who have kindly contributed their creativity, expertise, and time to make this anthology happen.

Introduction

One by Walking, Won by Walking

Roger Norum, Camilla Brudin Borg, Hanna Elisabet Åberg,
Susanne Österlund-Pötzsch, and John Martin

> My name is someone and anyone. I walk slowly, like one who comes from so far away he doesn't expect to arrive.
>
> —Jorge Luis Borges

Walking is having a moment. This might seem an odd thing to state, given how fundamental—quintessential, even—walking is to human activity. While humans are not the only bipeds bustling about the planet, we are certainly among its most present and active walkers. But walking is not merely the way we traffic the world. More than just a means of traversing space, of getting from point A to point B, walking is an embodied act of mobile connection (with the outside, with others, with ourselves) that deeply shapes who we are and how we perceive, engage with, and change the world.

In this brief introduction, we briskly walk readers through this innovative and fascinating transdisciplinary field of mobility research. We situate the volume within a broader intellectual lineage, exploring how walking and its historical and cultural meanings have been used as both a lens and a tool for research, and demonstrating how walking has been considered central to shaping perception and fostering reflection. Born from the mutual fascination and wonder of a small group of researchers and artists, this work reflects our ongoing thinking about what walking is, what it can do, and how it can help us mobilize a more sustainable future for the planet. This anthology emerges from years of walking and thinking together and constitutes the traces of walking together in practice—an embodied experience reformed into words.

While it is nothing new for scientists to study human movement, an increasing number of researchers and practitioners across multiple disciplines have begun to critically, creatively, and transdisciplinarily explore walking. More and more, walking is being used as a research and artistic tool, a philosophical practice, and a way of knowing that fosters attunement to landscapes, histories, and ecological concerns. Scholars in many fields are increasingly looking downward (at our feet) to discover what walking through the material world can teach us when we experiment with and engage in various forms of physical and conceptual mobility.

But what actually is walking? The definition is rather straightforward—and, well, pedestrian: to walk is to move at a regular pace by lifting and setting down each foot in turn, never having both feet off the ground at once. But while we may have a shared definition, the meanings and motions of walking are not the same for all: we walk and move with different techniques and in different ways—swiftly or with difficulty, for many reasons and at varying rates, and sometimes with different types of aids and devices. Walking, then, comprises a continuum of movement styles and necessities.

The interest in thinking, talking, and walking through all these issues was the starting point for founding the One by Walking research network in 2020. Organized through monthly online seminars and a series of intensive, place-based workshops across Sweden, Norway, Finland, and Iceland, the network has aimed to collaboratively advance innovative research on and through walking. It has grown into an active forum for discussion, experimentation, and cooperation, where we have sought to share and promote knowledge about the connections between bodies, landscapes, perception, movement, and sustainability—connections fueled by walking.

One by Walking was driven by the belief that thinking holistically and engaging with our connections to moving with and within space can ultimately lead to more sustainable environments and relationships with the Earth. We set out to build a methodological tool kit that promotes environmental and ecological awareness, inclusion, and connection to place and community. We believe that developing tools to better understand walking's multifarious relationships—and what walking can do for us as individuals, as a species, and as a planetary community—is an important step in addressing the critical environmental and social challenges currently facing the planet.

The nearly two dozen chapters in this anthology bring together contributions from almost as many fields across the social sciences, humanities, and artistic research, exploring the methodological and epistemological possibilities of walking. Focusing on embodied movement, sensory experience, and in situ enquiry, the chapters reflect on how walking itself—along with the theoretical and methodological approaches used to study it, and the worlds in which it is inscribed—can open up new ways of understanding mobility, sustainability, and the entanglements between bodies, landscapes, histories, and futures.

Walking has long influenced intellectual and artistic traditions. From the peripatetic philosophers of ancient Greece to Romantic poets such as William Wordsworth and naturalist-philosophers such as Henry David Thoreau, walking has been intimately linked to creativity, reflection, and epistemological inquiry. This tradition, in which walking engages with the surroundings, has become a means of inspiration and self-discovery. Walking always involves engagement with the environment, whether rural or urban. The flâneur—a quintessential figure of the modern city, theorized by Charles Baudelaire and Walter Benjamin—ambulated through fin-de-siècle urban spaces as both an observer and a participant, revealing the tensions between modernity and alienation (Coates 2017).

Walking has emerged as a powerful means of conducting research, one that amalgamates mobility, multisensory engagement, and the coproduction of knowledge between the moving researcher and the world (Bates et al. 2017). Qualitative sociologists, geographers, and anthropologists have long recognized the value of movement as a way to study place and social life. In the early twentieth century, classical ethnographers such as Bronisław Malinowski (1922) and Marcel Mauss (2008 [1934]) highlighted the significance of walking in shaping cultural practices. Later ethnographers further developed participatory and embodied approaches to fieldwork, incorporating movement as a key element of data collection. More recently, the rise of mobile research methods and walking ethnographies has underscored how walking fosters embodied understandings of space, memory, and social relations (Springgay and Truman 2018; O'Neill and Roberts 2019).

Tim Ingold's concept of "wayfaring" (2007), for example, positions walking as a mode of knowing that contrasts with static representations of space, emphasizing movement as a dynamic process through which people engage with their surroundings. Daniel Svensson et al. (2022) suggest thinking of walking and the paths inscribed in the landscape by repeated walking as "movement heritage." John Wylie's (2005) work demonstrates how bodily movement phenomenologically shapes the very experience of landscape. Indeed, walking methodologies have been central to various studies on rural and urban space, migration, and environmental change, offering a dynamic way to investigate human-environment interactions.

Following the late twentieth-century "spatial turn" in the social sciences, scholars shifted their focus to new ways of theorizing place and spatiality, shedding light on how ideas, people, and things move—or don't move—through space (Soja 1996). Their work advocated for novel methodological and theoretical approaches (Elliot et al. 2017) and introduced new considerations of how movement through space intersects with identity, power, and perception (Ahmed 2006).

The contemporary history of walking is closely tied to perceptions of and relationships between space and mobility, which evolve as the project of modernity descends on humankind. Many great thinkers have long seen walking as central to their practice: Henry David Thoreau's eager seclusion in Walden Woods, Gérard de Nerval's rambling to cure melancholy, Immanuel Kant's precisely timed marches through his hometown to escape the "compulsion of thought," and Friedrich Nietzsche's wandering the mountainside to write (Gros 2014). These philosophers all believed that the best creative thinking happens in solitude and that walking was essential for focusing their inner thought processes.

On the other hand, many creative writers and artists have valued walking for its ability to stimulate creative thought, reflection, and sensory attention, engaging both body and mind to generate a rhythm that facilitates new ways of thinking and perceiving. For example, Virginia Woolf described how walking shaped her literary imagination, providing her with a means to process emotions, think through complex ideas, and engage with the materiality of place in ways that static reflection simply could not (Larsson 2017). In her 1927 essay "Street Haunting," Woolf describes a walk across London to buy a pencil as an escape from the "shell-like covering which our souls have excreted to house themselves" (Woolf 2009). Yet Woolf's walks through London were far from solitary experiences. While wandering the streets as an "anonymous tramper," she locates in these moments "a central oyster of perceptiveness, an enormous eye," through which "one is not tethered to a single mind, but can put on briefly for a few minutes the bodies and minds of others" (Woolf 2009). Her urban walks are thus less about panoptic voyeurism and more about a shared vision of connectivity and compassion. In short, walking for Woolf is a process of empathetic cocitizenship with her fellow Londoners (Chacoff 2021).

In today's world, however, we cannot just sit (or walk) back and watch passively as the world changes. With the epoch of the Anthropocene now upon us—never mind what the International Union of Geological Sciences might have to say about the Earth's geological timetable—we are called to get up, stand up, move, and make others move. Walking takes on new significance as a mode of embedding humans in ecological awareness and sustainability thinking. While often framed as a low-impact, environmentally friendly form of transport, walking's role in fostering deeper ecological relationships goes well beyond carbon reduction. It can cultivate attentiveness to landscapes, rhythms, and environmental change, encouraging new forms of engagement—slow, fast, or otherwise—that resist extractive approaches to being in a place.

Environmental humanities scholars such as Anna Tsing have explored how walking fosters an ethics of care for fragile environments, highlighting how movement through environments is a decidedly relational act that shapes how humans interact with more-than-human worlds (Tsing 2015). Other important studies have explored bipedal mobilities using approaches from across various fields, including geography (Gandy 2024), tourism studies (Hall et al. 2017), folklore (Warneken 1989), ethnology (Sjöholm 2011), architecture and planning (Rosenberg 2012), literary studies (Larsson 2017), history (Solnit 2000), religion (Coleman and Eade 2004), anthropology (Ingold and Vergunst 2008), art and critical theory (Braidotti 2011; Sundberg 2014), and philosophy (Gros 2014), to mention just a few. While many disciplines are represented across much of this work, collaborative engagement between the humanities and social sciences, and with artists and other creative scholars, remains surprisingly rare—and ever more needed. In taking this very approach, we have found walking research to be particularly suited to experimental and arts-based approaches, as well as to bridging disciplinary divides. As a core human phenomenon and quotidian practice, walking is both a concrete activity and a semaphore for human relationships with the world around us. As such, it constitutes a fertile crossroads where multiple forms of scholarship can grow organically. The chapters included in this anthology suggest that walking can contribute to an innovative and generative methodological framework that is increasingly central to understanding and solving contemporary challenges.

Last, it is worth noting the collaborative nature of this book, as well as the network and workshops that spawned it. For Nietzsche, among many thinkers, movement helped to generate intellectual clarity and creativity. Walking afforded him the freedom to explore and play with thoughts and concepts, the luxury of being able to err or fail in his logics—without having to worry about others glimpsing the rawness of his thoughts. Nietzsche's extensive walking in the Swiss Alps was not merely a pastime but a vital part of his philosophical process, shaping his ideas on nature, will, and the body. On this practice, he wrote in *The Twilight of the Idols, or How to Philosophize with a Hammer* (Nietzsche,

1990 [1889]), "Alle wahrhaft großen Gedanken kommen einem beim Gehen." In English, this literally renders as "all truly great thoughts come to one while walking," though one common translation reads, "all great ideas are won by walking."

Borrowing directly from Nietzsche's words, the network adopted the moniker "Won by Walking" in its early days, emphatically suggesting the victorious epiphanies that can arise from consciously choosing to walk. But over the years, as our network grew, our work became more collaborative, and our walks became more entangled with one another. It became clear that "*One* by Walking" was a more apt title for who we were becoming (it didn't hurt that "won" and "one" are homonyms in English). While at first blush the "One" might seem to speak to the value of a walk in solitude, we feel rather that it highlights our practices of cocreating, cowriting, and cowalking, which were increasingly bringing us together as a One. We might then offer our own inspired, if liberal, translation of the Nietzsche quote: "All truly valuable knowledge is born as we become one by walking."

This book is the culmination of the conversations, interactions, and creative epiphanies—the wending and wielding, twisting and turning of walks and words—that have taken place in our network over the past four years. The contributions in this anthology engage with these various intellectual traditions, zooming in on how walking intersects with intellectual and cultural practices as diverse as ethnography, sensory perception, storytelling, cultural memory, and artistic creation. The chapters bring together historical and contemporary perspectives on walking practices to reveal how movement remains a vital mode of engaging with both past and present. They examine how pedestrian practices can contribute to environmental responsibility and ecological awareness. They explore how walking methodologies can serve both as research tools and as ethical engagements with place, focusing on the coproduction of knowledge through movement in fragile ecosystems and contested territories. They propose walking as a reflective and creative practice—one with the potential to foster relations and attachment to place, as well as to shape affects and atmospheres.

The chapters in this anthology are arranged into four themes. "Walking Reflections" showcases a trio of personal histories of walking that poetically capture everyday landscapes of past and present. Next, the chapters in "Hands-on/Foot-Out Walkshops and Workshops" speak to various ways of walking and means of studying walking, showing the range of cogitation and creativity that arises when somewhere new presents itself to be walked. "Engaging Walking" presents five pieces of writing that ponder interactive and cocreated walking in particular environments, entangled with the surface of the Earth. Finally, in "Methodologies Underfoot," seven chapters sound out echoes of walking as method—practiced, swayed, interrupted, regained—describing various uses of mobility and immobility as practices for better understanding the world. As an entire body of work, this volume is born out of collaboration—out of walking and speaking and thinking and exploring and experimenting together. It is a testament to our belief that walking, whether alone or with another, can open new pathways for understanding our relationships to place, movement, knowledge—and, indeed, to ourselves.

Figure 0.1 Created by Dan Borg.

References

Ahmed, Sara. 2006. *Queer Phenomenology: Orientations, Objects, Others.* Duke University Press.

Bates, Charlotte, and Alex Rhys-Taylor, eds. 2017. *Walking through Social Research.* Routledge.

Borges, Jorge Luis. 2012 [1925]. "Jactancia de quietud." In *Poesía complete / Complete Poetry.* Vintage Español.

Braidotti, Rose. 2011. *Nomadic Theory: The Portable Braidotti.* Columbia University Press.

Chacoff, Alejandro. 2021. "Is the Digital Age Costing Us Our Ability to Wander?" *New Yorker*, August 23. https://www.newyorker.com/magazine/2021/08/30/is-the-digital-age-costing-us-our-ability-to-wander.

Coates, Jamie 2017. "Key Figure of Mobility: The Flâneur." *Social Anthropology* 25 (1): 28–41.

Coleman, Simon, and John Eade, eds. 2004. *Reframing Pilgrimage: Cultures in Motion.* Routledge.

Elliot, Alice, Roger Norum, and Noel B. Salazar. 2017. *Methodologies of Mobility: Ethnography and Experiment.* Berghahn Books.

Gandy, Matthew. 2024. "Attentive Observation: Walking, Listening, Staying Put." *Annals of the American Association of Geographers* 114 (7): 1386–404. https://doi.org/10.1080/24694452.2024.2353841.

Gros, Frédéric. 2014. *A Philosophy of Walking.* Translated by John Howe. Verso.

Hall, Michael, Yael Ram, and Noam Shoval, eds. 2017. *The Routledge International Handbook of Walking.* Taylor and Francis.

Ingold, Tim. 2007. *Lines. A Brief History.* Routledge.

Ingold, Tim, and Jo Lee Vergunst, eds. 2008. *Ways of Walking: Ethnography and Practice on Foot.* Anthropological Studies of Creativity and Perception. Ashgate.

Larsson, Lisbeth. 2017. *Walking Virginia Woolf's London: An Investigation in Literary Geography.* Palgrave Macmillan.

Malinowski, Bronislaw. 1922. *Argonauts of the Western Pacific.* Routledge and Kegan Paul.

Mauss, Marcel. 2008 [1934]. "The Techniques of the Body." In *The Design Culture Reader*, edited by Ben Highmore. Routledge.

Nietzsche, Friedrich. 1990 [1889]. *The Twilight of the Idols and the Anti-Christ, or How to Philosophize with a Hammer.* Translated by R. J. Hollingdale. Penguin Classics.

O'Neill, Maggie, and Brian Roberts. 2019. *Walking Methods. Research on the Move.* Routledge.

Rosenberg, Elissa. 2012. "Walking in the City: Memory and Place." *The Journal of Architecture* 17 (1): 131–49. https://doi.org/10.1080/13602365.2012.659914.

Sjöholm, Carina. 2011. *Litterära resor: turism i spåren efter böcker, filmer och författare.* Makadam.

Soja, Edward. 1996. *Thirdspace: Journeys to Los Angeles and Other Real-and-Imagined Places.* Blackwell.

Solnit, Rebecca. 2000. *Wanderlust: A History of Walking.* Viking.

Springgay, Stephanie, and Sarah E. Truman. 2018. *Walking Methodologies in a More-than-Human World.* Routledge.

Sundberg 2014. "Decolonizing Posthumanist Geographies." *Cultural Geographies* 21 (1): 33–47.

Svensson, Daniel, Katarina Saltzman, and Sverker Sörlin. 2022. *Pathways: Exploring the Routes of a Movement Heritage.* White Horse Press.

Tsing, Anna. 2015. *The Mushroom at the End of the World: On the Possibility of Life in Capitalist Ruins.* Princeton University Press.

Warneken, Bernd Jürgen. 1989. "Bürgerliche Gehkultur in der Epoche der Französischen Revolution." *Zeitschrift für Volkskunde* 85: 177–87.

Woolf, Virginia. 2009 [1927]. "Street Haunting: A London Adventure." In *Selected Essays*. Oxford University Press.

Wylie, John. 2005. "A Single Day's Walking: Narrating Self and Landscape on the South West Coast Path." *Transactions of the Institute of British Geographers* 30 (2): 234–47.

1

Whose Healing? Reflexive Walking and a More-than-Human Perspective on Moving-with-Nature

Noel B. Salazar

The practice of walking is as old as humankind. What's more, according to scholars from various scientific disciplines (DeSilva 2021), walking is what made us human in the first place. Throughout the ages, a multiplicity of meanings has been attributed to walking practices. Walking has served many different purposes, and various (ideal) "types" of walkers have been described. One of the first essays specifically on walking was William Hazlitt's (1822) *On Going a Journey*, which established the parameters for walking in nature. When thinking about walking, we often envision it as a human activity—something people do somewhere, with the environment in which the walking takes place being a mere passive background. This anthropocentric understanding of walking is typically Western. At the same time, there is a growing acknowledgment that "walking makes us aware of, and allows us to flourish as, the relational beings we already are" (Coeckelbergh 2015, 141). Tim Ingold has long argued that such a relational way of thinking involves "treating the organism not as a discrete, prespecified entity but as a particular locus of growth and development within a continuous field of relationships" (2004, 219).

In a similar vein, David Macauley muses: "In the act of walking, we are introduced to and into a larger order, a more capacious natural and social world. Through our ambling bodies, we can discover an elemental relationship to the earth, a robust, even when culturally mediated, processual field of phenomena flowing beneath—but not only beneath—our fleeting feet" (Macauley 2001, 15). By restoring us to all of our senses and thereby placing us relationally in the world, walking offers an "epistemology of contact" (18). One could argue that walking is an activity that requires an elevated level of cognitive ability. However, it should be noted that "knowing" is not solely confined to the mind but rather is dispersed throughout the entire network of embodied connections formed by the presence of beings on this planet.

While geologically speaking, we currently live in the Cenozoic Era, Thomas Berry (in Swimme and Berry 1992) proposed a future Ecozoic Era, defined by the recognition that human life is only one part of the greater living planetary processes that hold and sustain us. The term captures well the integral functioning of life systems in their mutually enhancing relations. Already in the 1950s, Robert Redfield (1953) argued that we cannot talk about humans and nature separately because humans are a part of nature. If we have a sense that we are "all planetary life," then the inseparable connection between life and nonlife shows that we must broaden our self-definition [or definition of self] to include the nonliving components of the planet. The fields of posthuman ecocriticism and new materialism take such a comprehensive approach by focusing on entities such as animals, water, stones, forests, and the planet itself, ascribing qualities like force, thought, agency, or emergence to them (Oppermann 2016; Bray, Eaton, and Bauman 2023; see also chapter 16 in this anthology).

In general, more-than-human approaches help us deconstruct dominant dualistic ideologies (Welz 2021). They point to the ethical and political relevance of moving-with (the environment) to feel and think through the intensities of the entanglements of human and world (Springgay and Truman 2018). Moving-with forces people to rely on their affective experiences. Becoming accountable to the more-than-human involves acknowledging the erasures of other knowledges and methods—erasures that, in part, have framed thinking about the more-than-human as only a recent turn in thought. Moving-with, then, becomes not only a movement of thought with others but also a process of engaging with erased or disavowed histories. As such, it is a dynamic reengagement with the ways in which the planet and its constituent parts have been understood, protected, feared, and treasured.

Indeed, while more-than-human epistemologies may seem novel from a contemporary Western perspective, Indigenous knowledge systems show a different picture. The "topokinetic" (Turnbull 2007) nature of knowing-through-movement has been described on every continent (Green 2020). In diverse cultural contexts, people establish their sense of identity through the deliberate act of walking and experiencing their own physical movement within a landscape rich with narratives and histories. The Indigenous peoples of North America, for example, express the concept of living in harmony with the natural world by referring to it as "walking in balance on the Earth" (Hogan 2009; Johnston 2014), where every stride is regarded as a respectful salutation to the entity known as Mother Earth. A state of overall well-being and a sense of belonging in one's surroundings are outcomes of regular realignment and synchronization with the fundamental forces of nature, achieved through the practice of rituals. The initiatory "walkabout" undertaken by Aboriginal Australians, for instance, is an act in which people turn toward the more-than-human for understandings that vitalize and sustain the human community (Peterson 2004). According to tradition, this is done by periodically making a ritual journey along the Dreaming track, walking in the footsteps of the clan Ancestor. Australian Aboriginal people sing out loud when they walk in the wild, talking to the rest of the natural world as if it were family.

In such Indigenous frameworks, protecting the environment does not come at the expense of human prosperity and well-being; it is the source of both. We are an integrated living system in constant motion, surrounded by and infused within a large entity called the biosphere, which includes all living beings. Unfortunately, many people are unaware that the vast, expansive living system that inhabits our planet is who "we" are. Moreover, "as a result of [the] shift from stewardship and trust to ownership and control it is harder for us to behave as people with commitments to the wider Earth community's health and well-being" (Higgins 2012, 107). Ironically, imagining anthropogenic climate change and its associated ecological and sociocultural transformations has opened up possibilities for developing new types of human-environment relations. In the act of slow and reflexive walking, for example, people are (re) introduced to and into a larger order, a more capacious natural and social world. A phenomenological approach to our "human-being-in-the-world" attempts to overcome the Cartesian, dualistic separation of body and mind, nature and culture, that still permeates Western societies and sciences. In such an integrated approach, we do not consider nature and other environments from a distance but rather as something of which we are an integral part. This chapter, then, reflects on the importance of walking for the human-environment relationship and the possible tensions and contradictions this may entail.

Wild Imaginaries

According to the posthuman paradigm, humans are living tissue composed of diverse aspects of the world, which momentarily make up the human body before transforming into other systems. A lifeworld is characterized by the fact that humans cannot be separated from the dynamic environment in which they live and are immersed (e.g., Donna Haraway's [2016] concept of *sympoiesis* or "making-with"). Contemporary attempts to (re)connect humanity to the (natural) environment are taking multiple forms, many of which can be linked historically to well-established but dated ideas of romanticism in Europe and transcendentalism in the United States (as well as the incorporation of Asian spiritual influences in both philosophies). In my own research on recreational outdoor practices such as hiking and running, for instance, traces of these connections are evident in nature-based walking meditation, forest bathing (*shinrin-yoku*), chi-running (inspired by tai chi), and "yogging" (a combination of yoga and running) (Salazar 2018b).

In Western societies, notions of the wild and wilderness have long been understood to refer to empty spaces devoid of culture, signs of civilization and development, and human presence. In the sixteenth and seventeenth centuries, Europeans laid the intellectual foundation for a more favorable attitude toward so-called wild nature. The notions of the sublime and picturesque played a pivotal role in promoting the aesthetic appreciation of untamed landscapes. Combined with the imaginary of a life closer to nature, these concepts contributed to the development of the Romantic movement. Romantic writers and poets fostered a taste for nature, reinforcing the idea that the proper way to commune with nature was by walking through the countryside. In this tradition, "walking is supposed to be 'wholesome' in the sense that it heals you from modern alienation and unites you with nature" (Coeckelbergh 2015, 138). Yet, as critics have pointed out, these ideas are nothing more than imaginaries that have regularly resulted in the erasure of Others (e.g., Indigenous people) and in the blind reinforcement of an artificial nature-culture binary (Vannini and Vannini 2016).

Despite scholarly criticisms, since the mid-twentieth century, there has been a considerable surge in the demand for experiencing "nature." Outdoor experiences in natural environments have become immensely popular. With increasing leisure time and mobility, and through the establishment of national parks and wilderness areas, the natural environment is now accessible to large numbers of people. However, the idea that wilderness or outdoor recreation can be supplied in this way seems at odds with the notion of wilderness as a space of implied freedom. The specific form of nature we wish to, should, or need to (re)establish a connection with, as well as the method of engaging with it, may vary depending on the motivations behind seeking such a relationship. If our motivation is merely to alleviate stress, a walk in an urban park should suffice. Elizabeth Willott nicely captures the tension between how nature is imagined and the kind of nature people want for themselves: "When we say we desire to experience nature (including restored environments), what most of us mean is that we desire to experience selected aspects of nature. We prefer not to experience serious illness or traumatic death. [. . .] While recognizing the importance of predators in an ecosystem, we prefer that our families are not the prey" (2004, 148–49). This focus on nature as something clearly beneficial, excluding problematic wildlife, follows many past attempts to bring humans and nature "together." Such efforts, epitomized in the tradition of Arcadia—a vision of a lost form of living in harmony with nature dating back to ancient Greek mythology—often involved a specific idea of wildness. Over the centuries, the Arcadian tradition has taken two main forms: the sublime wilderness (or primitive) and the rural idyll (or pastoral).

By arguing that meaning is invariably "made" by sociocultural systems, as is commonly done in the social sciences and humanities, the more-than-human world is stripped of agency and valency. Nature is elided, rendered unable to participate in its own representations. Constructivism sought to give voice to minorities who had been repressed by the structuring of discourse—but in the case of the natural world, constructivism left the Other of animal, plant, and ecosystem silenced. Because, according to constructivism, there is no longer any "nature" or "wild," and because we have all accepted the culturally constructed and always already artefactual status of such designations, there is nothing to stop us from technologically remaking our landscapes without reference to such sentimental antiquarianisms as "biodiversity," "ecological baselines," or "uniqueness." For some time now, those advocating for the creation or preservation of significant space for more-than-human nature—or what the American Wilderness Act calls "the earth and its community of life" (US Congress 1964, 891)—have found themselves quickly denounced as romantic or nostalgic.

In any case, "wild" places around the globe continue to exert a powerful pull on the popular imagination. As Robert Macfarlane (2007) puts it, it is less about the clear absence of humans and their history and culture from these places and much more about the subjective feeling of being alive in a place animated by at least a modicum of self-will (the idea at the etymological root of the word "wild"), on which wildness—as a process rich with vitality and unpredictability—depends. For Macfarlane, walkers leave their mark on the landscape and navigate it in two complementary ways. On a literal, geomorphological level, as we walk, we leave a trace on the ground with each step. However, as poets and writers have suggested for centuries, the paths we abrade into the earth also leave an imprint on us, pressing themselves into our imagination. We carry these imprints in our minds as we make our way in the world. When we walk, land and landscape also leave their mark on our bodies. Walking attunes us to the connection between humans and the more-than-human, to the physical and chemical reactions that occur when we walk through a place. Our bodies are constantly entangled with the weather, the environment, and the air we breathe. These things circulate through and within us as we move, which brings us to questions of methodology.

Walking as Method

Walking has been widely adopted as an embodied methodological practice (Lorimer 2011). In *Ways of Walking: Ethnography and Practice on Foot*, Tim Ingold and Jo Vergunst (2008) present various studies, papers, and collective efforts to conceptualize walking as an essential human experience, social activity, and way of thinking and (re)searching. Their phenomenological orientation places walking at the heart of how humans come to be in and know the world they inhabit. Insights into how people walk, create routes, and reflect on these practices offer a key to understanding their ways of being in the world. The scholarship on walking examines how the act of walking broadens the scope of inquiry to include aspects such as location, sound, and interaction—elements that embody the spontaneous and evolving nature of engaging with the world (Flint 2022). Walking methodology is particularly suitable for relational, material, and situated research (Springgay and Truman 2018, 14).

Hayden Lorimer argues that "in the midst of walking it is possible to think about being-in-the-world, and find grounds for a freer experiential exploration of what it is *to be* or *become*" (2011, 23). In this perspective, the walker seeks to reach into the "intimacies of encounter—rather than focusing on the outer (symbolic or socialized) meaning of walking arts" (23). One example is John Wylie's documented walk along the South West Coast Path, in which he sought "to activate a space and time within which [he] might engage with and explore issues of landscape, subjectivity and corporeality" (Wylie 2005, 234). Such reflexive walking is a practice that involves walking with a heightened awareness of one's surroundings and bodily sensations. It is a form of "attentive walking"

(Lilja 2022) that encourages individuals to be present in the moment and to reflect on their thoughts and feelings as they walk. This combination of the reflexive walking self in the midst of shifting hues, tones, and colors and the "intensity of places and situations" (Wylie 2005, 236) generates affect—"an intensity, a field, perhaps, of awe, irritation or serenity"—which brings together walker and landscape in a union where it becomes impossible to tell one from the other: "landscape and self are mutually configured" (239).

In other words, walking reflexively becomes a walking-with multiple Others (Salmela and Valtonen 2019), "inviting rhythmic, temporal and affective dimensions to our social-scientific inquiry" (Rantala, Valtonen, and Salmela 2020, 40). The idea of walking as a multisensory, embodied experience has been increasingly discussed across the social sciences and humanities (Pink et al. 2010; Salazar 2024). As David Macauley notes, "Walking puts the world back into perspective. Via our porous and sensing bodies, walking deepens and thickens the perceptual scene, unlike seated driving which tends to thin, draw out and attenuate our experiences" (2001, 26). That the senses interrupt and mediate everyday walking practices is something Indigenous peoples have long recognized: "A man in Anaktuvuk Pass, in response to a question about what he did when he visited a new place, said to me, 'I listen.' That's all. I listen, he meant, to what the land is saying. I walk around in it and strain my senses in appreciation of it for a long time before I, myself, ever speak a word" (Lopez 1986, 230).

Reflexive walking can thus be understood as a kind of embodied experiential learning. Karein Goertz elucidates how walking promotes divergent (rather than convergent) thinking, noting that it "is an excellent facilitator of original, innovative thinking because the physical act of left-right steps simulates the integration of the two halves of the brain, the logical and the intuitive" (2018, 61). As such, walking is an interactive way of knowing or coming-to-know, allowing the entire body and all of its senses to experience the surroundings, trace and connect different areas, intuitively sense when and how to avoid potential dangers, and live within entangled social pathways (Jung 2014, 625). In reflexive walking (or what authors such as Yuha Jung [2014] call "mindful walking"), the body is engaged in the act of walking, which then becomes the lived experience: information is received in multiple ways—seeing, hearing, touching, smelling, tasting, overall sensing, and mentally connecting. While walking in the same environment, different individuals are attuned to different stimuli, each navigating a sensory universe tied to their personal histories.

Within walking scholarship, the individual human (often White, privileged, and male) bipedal body has dominated as the figure of movement, with environments, organisms, and other people often relegated to the role of scenic backdrop (Springgay and Truman 2018). One possible new thread of research would be to explore different practices of how people walk-with environments and how environments walk-with them. This requires a certain "more-than-human intimacy," referring to "the sensual, emotional, spiritual, and personal entanglements" that walking involves "in a constant and thorough proximity" (Einarsdóttir and Lund 2024, 107). Another thread would be to consider how decentering the human can generate diverse ways of understanding environmental movements. The point here is not to stretch walking into an amorphous analytic but to consider it as a cascade of relations and relationalities through which entities and worlds mobilize and are mobilized (Sundberg 2014), leading, among other things, to "more-than-human care" (Rantala, Valtonen, and Salmela 2020). These relations, moreover, can be activated by decentering humans as situated within specific places. Such a realignment can draw attention to ways of knowing and being in sites that also overlook other environment inhabitations—from Indigenous experiences and cosmologies to more-than-human relations—that are often rendered less significant when addressing environmental change.

Natasha Myers (2016) has experimented with walking to think-with plant sentience. She contends that methods are practices for cultivating "*modes of attention* that might help *tune in* to the deep time of these lands and the *natural-cultural happenings* shaping its present" (85). These modes of attention inquire into the significance of many aspects pertaining to the land. Such practices of attunement allow researchers to become more consciously aware of the "qualities, rhythms, forces, relations and movements" (Stewart 2011, 445) of the humans and nonhumans they are researching with. The act of tuning into something is distinct from the acts of capturing or documenting it. It involves bending and folding, or what Carla Hustak and Natasha Myers (2012) call "involutionary momentum." Involution, as opposed to evolution, is a relational movement—a coupling or comingling. Involutionary readings, they write, "give way to livelier ontologies and intra-active worldings" (105). The focus lies on the exploration of uncertainties, prompting the question of whether methods of physical presence, cognitive focus, and creative thinking are necessary to get a comprehensive understanding of a location. As Juanita Sundberg (2014) notes, walking-with is a way of walking toward more pluriversal worlds and encounters.

It is important to note that Indigenous scholars have critically interrogated the more-than-human turn, arguing that it continues to erase Indigenous understandings that have always attended to nonhuman animacy (Todd 2016). The (predominantly Western) scholars to whom we commonly attribute the remarkable understandings of concepts like "more-than-human" sentience and agency are not the actors responsible for constructing and upholding the knowledge systems that European and North American anthropologists and philosophers have extensively examined for over a century. These knowledge systems, Indigenous in origin, have formed the foundation upon which many contemporary ontological breakthroughs and reconceptualizations are based. By exclusively referencing Western intellectuals who engage with the concept of the "more-than-human," while neglecting to acknowledge the contributions of Indigenous scholars who explore similar themes, we inadvertently reinforce the hegemony of Western knowledge.

Healing Walks

The experience of nature has often been believed to have therapeutic effects and healing properties, with significant theoretical and empirical data to support this claim. Empirical support for the value of contact with nature is growing at an enormous rate, with an extraordinary degree of consistency across research paradigms and theoretical views. Studies have shown that wilderness has a restorative quality, that outdoor physical exercise has a positive impact, and that people generally enjoy the challenge of being in the outdoors. Henry David Thoreau (1817–1862), one of the forebears of the environmental movement, was among those who included nature as an element of (the socioculturally constructed notion of) the good life. As Thoreau (1862) helped to reveal, walking in the wilderness is guided by—and often directed through—the encompassing four classical elements of earth, air, fire, and water. Walking, then, is a way of becoming attuned to the earth and other elements, especially—but not only—in a wilderness setting. Interestingly, Thoreau virtually ignored contemporary environmental concerns about the preservation of habitats and species (Turner 1996, 82).

The idea of nature experiences being therapeutic and healing for the individual in a time of climate and ecological breakdown deserves scrutiny and reflection. While those who follow in Thoreau's footsteps obviously no longer ignore the issues of environmental degradation, they regard nature as an essential part of a good life and see its rehabilitation as integral to our own rehabilitation. Ironically, the very source of well-being and health benefits is the same resource and commodity that is being systematically destroyed by our ecocidal, carbon-fueled societies—societies that continue to fail to make the radical and urgent adaptations required for new, environmentally sustainable ways of living.

When Aldo Leopold referred to the "health" of the land, it was less metaphorical than one might think. He defined health as the "capacity of the land for self-renewal" and ecological conservation as "our effort to understand and preserve this capacity." The individual, acting within the framework of a land ethic and guided by an ecological conscience, should take "responsibility for the health of the land" (Leopold 1949, 258). Leopold's prophetic message was that the renewal of the land's health would be a renewal of society's health. Similarly, ecopsychology argues that, because we are naturally inclined to connect with and relate to our planet and the natural world, we will psychologically (and materially) suffer if our planet falls deeper into environmental chaos. The modern health crisis and the deterioration of planetary health stem from severed connections with nature (Zelenski et al. 2023). Nature-based solutions are proposed to restore health across the personal-to-planetary health continuum. While walking, alongside other "healthy mobilities" (Bell and Cook 2021) such as cycling, is often celebrated as a "best buy for public and planetary health" (Bull and Hardman 2018), it is important to acknowledge that these practices may also negatively impact planetary health. The coronavirus crisis made us acutely aware of this tension (Salazar 2021).

The recognition of the damaging impact of human activity on nature has led to the theory and practice of rewilding, a progressive approach to conservation biology developed in the late 1980s. Rewilding involves letting nature take care of itself, enabling natural processes to shape land and sea, repair damaged ecosystems (e.g., by reintroducing species to environments where they were once present before humans drove them off), and restore degraded landscapes (Hawkins et al. 2023). Through rewilding, wildlife's natural rhythms create wilder, more biodiverse habitats and restore connectivity between fragmented protected areas. Paradoxically, as processes such as rewilding and restoration help nature become "wilder" again, they also make nature even more attractive for people to explore through walking and other activities in the sphere of leisure and tourism. Many places around the world are struggling to address this tension.

Conclusion

Through dominant sociocultural practices, we are persistently taught how to read and react to nature rather than allowing ourselves to understand and come to know nature through our own active embodiments and movements. We are human and more-than-human, tangible and intangible, and everything in between—especially in those relationships that constantly bring us into being. As humans, we are perpetually engaged in a state of coexistence with entities that are not of human nature and are never situated outside of a complex network of interconnections or an ecological system. To engage with the more-than-human is to acknowledge that we are not separate from other species, nor from our shared precarious futures. In this context, "Walking is ultimately a responsibility, a 'responding-to' the coexistence intrinsic to place, what we have been describing as a 'call' that asks we offer something of ourselves to the human and nonhuman others whose relationships support us" (Larsen, Johnson, and Wildcat 2017, 48).

Amid the prevailing sense of contemporary alienation, particularly among urbanites, the act of walking emerges as a subversive endeavor, enabling people to contemplate the intricate interplay between their physical selves and the surrounding environment (Salazar 2022). Walking is a foundational more-than-human practice and very likely one of the reasons people feel so much at home while moving-with nature and derive so much pleasure from it. The act of walking, considered a fundamental human activity, has the ability to foster a sense of groundedness and restore a feeling of belonging with the natural environment. No one walks, behaves, senses, or interprets the same way. This is what makes walking so special in the most basic sense of the word. The act of walking takes on a highly significant, indeed pivotal, role, becoming the means through which human beings learn to understand the world around them as they pass through it. The trace they leave behind is not merely recorded in the paths left in their wake but also in the oral histories and texts through which such actions are recorded (Salazar 2018a).

There is an existential component to walking, where this most natural and instinctive act becomes, at the same time, a conscious activity. As Tim Ingold and Jo Vergunst (2008, 2) state, "walking is not just what a body *does*; it is what a body *is*." The capacity of walking to alter consciousness is hugely significant because it opens possibilities for encounters with landscapes that engage with the beyond-human and the supranatural. If one walks in this reflexive way, strange things may always happen because walking opens us to those possibilities. Extensive scholarly inquiry has been conducted on walking, examining the influence of psychological, social, and environmental factors on individuals' engagement in physical activity. Furthermore, it has acknowledged the positive effects of physical activity on mental health and overall well-being (Zurawik 2020). Physicians in various countries are now prescribing nature or "green prescriptions" to their patients (Bell et al. 2019). Yet the human-landscape relations generated by walking also have the potential to expose vulnerabilities.

Walking is believed to have lower environmental impacts than other activities in natural environments. However, there is a tension between the need for individual connection with "nature" and the negative environmental impact this often entails. Consider large-scale erosion, damaged vegetation, disruption to wildlife, altered hydrology, widening trails, and increased muddiness. Walking as a means to an end becomes walking as an end in itself, a form of therapy (Gatrell 2013)—but only to heal the human. In other words, "walking in 'nature' is not necessarily good; other activities may be environmentally better and, if we want to walk at all, what matters is *how* we walk" (Coeckelbergh 2015, 8–9). We need to move beyond the category of "nature" to recognize how experiences of health and well-being emerge through dynamic, more-than-human relations, such as with individual species, creatures, flows of weather, terrain, technology, and tools (Bell et al. 2019). The concept of health must be reinvented as a dynamic entity arising from intricate linkages and connections that include not only the individual physical body but also the more-than-human. These shifting relational configurations continually (re)shape people's capacities to affect and be affected in ways that produce moments of health and well-being or, conversely, experiences of illness, impairment, or exclusion. After all, healing means nothing more than making whole again, thus requiring a holistic approach.

References

Bell, Sarah L., and Simon Cook. 2021. "Healthy Mobilities." *Transfers* 11 (2): 98–108.

Bell, Sarah L., Catherine Leyshon, Ronan Foley, and Robin A. Kearns. 2019. "The 'Healthy Dose' of Nature: A Cautionary Tale." *Geography Compass* 13 (1): e12415.

Bray, Karen, Heather Eaton, and Whitney Bauman, eds. 2023. *Earthly Things: Immanence, New Materialisms, and Planetary Thinking*. Fordham University Press.

Bull, Fiona C., and Adrianne E. Hardman. 2018. "Walking: A Best Buy for Public and Planetary Health." *British Journal of Sports Medicine* 52 (12): 755–56.

Coeckelbergh, Mark. 2015. *Environmental Skill: Motivation, Knowledge, and the Possibility of a Non-Romantic Environmental Ethics*. Routledge.

DeSilva, Jeremy. 2021. *First Steps: How Walking Upright Made Us Human*. Harper.

Einarsdóttir, Elva Björg, and Katrín Anna Lund. 2024. "Walking-with Landscape." In *Researching with Proximity: Relational Methodologies for the Anthropocene*, edited by Outi Rantala, Veera Kinnunen, and Emily Höckert. Springer.

Flint, Maureen A. 2022. "More-than-Human Methodologies in Qualitative Research: Listening to the Leafblower." *Qualitative Research* 22 (4): 521–41.

Gatrell, Anthony C. 2013. "Therapeutic Mobilities: Walking and 'Steps' to Wellbeing and Health." *Health & Place* 22: 98–106.

Goertz, Karein K. 2018. "Walking as Pedagogy." In *The Routledge International Handbook of Walking*, edited by Colin Michael Hall, Yael Ram, and Noam Shoval. Routledge.

Green, Lesley. 2020. *Rock / Water / Life: Ecology & Humanities for a Decolonial South Africa*. Duke University Press.

Haraway, Donna Jeanne. 2016. *Staying with the Trouble: Making Kin in the Chthulucene*. Duke University Press.

Hawkins, Sally, Ian Convery, Steve Carver, and Rene Beyers, eds. 2023. *Routledge Handbook of Rewilding*. Routledge.

Hazlitt, William. 1822. "On Going a Journey." *New Monthly Magazine* 4: 73.

Higgins, Polly. 2012. *Earth Is Our Business: Changing the Rules of the Game*. Shepheard-Walwyn Publishers.

Hogan, Linda, ed. 2009. *The Inner Journey: Views from Native Traditions*. Parabola anthology series. Morning Light Press.

Hustak, Carla, and Natasha Myers. 2012. "Involutionary Momentum: Affective Ecologies and the Sciences of Plant/Insect Encounters." *differences* 23 (3): 74–118.

Ingold, Tim. 2004. "Beyond Biology and Culture: The Meaning of Evolution in a Relational World." *Social Anthropology* 12 (2): 209–21.

Ingold, Tim, and Jo Lee Vergunst, eds. 2008. *Ways of Walking: Ethnography and Practice on Foot*. Anthropological Studies of Creativity and Perception. Ashgate.

Johnston, Basil. 2014. *Walking in Balance: Meeyau-ossaewin*. Kegedonce Press.

Jung, Yuha. 2014. "Mindful Walking: The Serendipitous Journey of Community-Based Ethnography." *Qualitative Inquiry* 20 (5): 621–27.

Larsen, Soren C., Jay T. Johnson, and Daniel R. Wildcat. 2017. *Being Together in Place: Indigenous Coexistence in a More than Human World*. University of Minnesota Press.

Leopold, Aldo. 1949. *A Sand County Almanac: And Sketches Here and There*. Oxford University Press.

Lilja, Petra. 2022. "Attentive Walking: Encountering Mineralness." In *Pathways: Exploring the Routes of a Movement Heritage*, edited by Daniel Svensson, Katarina Saltzman, and Sverker Sörlin, 201–18. The White Horse Press.

Lopez, Barry Holstun. 1986. *Arctic Dreams: Imagination and Desire in a Northern Landscape*. Scribner.

Lorimer, Hayden. 2011. "Walking: New Forms and Spaces for Studies of Pedestrianism." In *Geographies of Mobilities: Practices, Spaces, Subjects*, edited by Tim Cresswell and Peter Merriman. Ashgate.

Macauley, David. 2001. "Walking the Elemental Earth: Phenomenological and Literary 'Foot Notes.'" In *Passions of the Earth in Human Existence, Creativity, and Literature*, edited by Anna-Teresa Tymieniecka. Kluwer Academic Publishers.

Macfarlane, Robert. 2007. *The Wild Places*. Granta.

Myers, Natasha. 2016. "Becoming Sensor in Sentient Worlds: A More-than-Natural History of a Black Oak Savannah." In *Anthropology of the Arts: A Reader*, edited by Gretchen Bakke and Marina Peterson. Routledge.

Oppermann, Serpil. 2016. "From Posthumanism to Posthuman Ecocriticism." *Relations Beyond Anthropocentrism* 4 (1): 24–37.

Peterson, Nicolas. 2004. "Myth of the 'Walkabout': Movement in the Aboriginal Domain." In *Population Mobility and Indigenous Peoples in Australasia and North America*, edited by John Taylor and Martin Bell. Routledge.

Pink, Sarah, Phil Hubbard, Maggie O'Neill, and Alan Radley. 2010. "Walking across Disciplines: From Ethnography to Arts Practice." *Visual Studies* 25 (1): 1–7.

Rantala, Outi, Anu Valtonen, and Tarja Salmela. 2020. "Walking with Rocks—With Care." In *Ethics and Politics of Space for the Anthropocene*, edited by Anu Valtonen, Outi Rantala, and Paolo Davide Farah. Edward Elgar Publishing.

Redfield, Robert. 1953. *The Primitive World and Its Transformations*. Cornell University Press.

Salazar, Noel B. 2018a. *Momentous Mobilities: Anthropological Musings on the Meanings of Travel*. Berghahn.

Salazar, Noel B. 2018b. "'Moveo ergo sum': Mobility as Vital to Humanity and Its (Self) Image." In *The Human Image in a Changing World: Proceedings of the 5th World Humanities Forum*. National Research Foundation of Korea.

Salazar, Noel B. 2021. "Existential vs. Essential Mobilities: Insights from Before, During and After a Crisis." *Mobilities* 16 (1): 20–34.

Salazar, Noel B. 2022. "The Art of Urban Walking." In *Teaching Urban Art Practices in Pandemic Times*, edited by Sarah Caillard. Blurbs.

Salazar, Noel B. 2024. "Afterthoughts on Multisensoriality and the Heritage-Mobility Nexus." In *Cultural Heritage and Mobility from a Multisensory Perspective*, edited by Magdalena Banaszkiewicz and Karolina Nikielska-Sekula. Routledge.

Salmela, Tarja, and Anu Valtonen. 2019. "Towards Collective Ways of Knowing in the Anthropocene: Walking-with Multiple Others." *Matkailututkimus* 15 (2): 18–32.

Springgay, Stephanie, and Sarah E. Truman. 2018. *Walking Methodologies in a More-than-Human World: WalkingLab*. Routledge.
Stewart, Kathleen. 2011. "Atmospheric Attunements." *Environment and Planning D: Society and Space* 29 (3): 445–53.
Sundberg, Juanita. 2014. "Decolonizing Posthumanist Geographies." *Cultural Geographies* 21 (1): 33–47.
Swimme, Brian, and Thomas Berry. 1992. *The Universe Story: From the Primordial Flaring Forth to the Ecozoic Era—A Celebration of the Unfolding of the Cosmos*. Harper.
Thoreau, Henry David. 1862. "Walking." *The Atlantic Monthly* 9 (56): 657–74.
Todd, Zoe. 2016. "An Indigenous Feminist's Take on the Ontological Turn: 'Ontology' Is Just Another Word for Colonialism." *Journal of Historical Sociology* 29 (1): 4–22.
Turnbull, David. 2007. "Maps Narratives and Trails: Performativity, Hodology and Distributed Knowledges in Complex Adaptive Systems—An Approach to Emergent Mapping." *Geographical Research* 45 (2): 140–49.
Turner, Jack. 1996. *The Abstract Wild*. University of Arizona Press.
US Congress. 1964. An Act to Establish a National Wilderness Preservation System for the Permanent Good of the Whole People, and for Other Purposes. 78 Stat. 890. US Government Printing Office.
Vannini, Phillip, and April Vannini. 2016. *Wilderness*. Routledge.
Welz, Gisela. 2021. "More-than-Human Futures: Towards a Relational Anthropology in/of the Anthropocene." *Hamburger Journal für Kulturanthropologie* 13: 36–46.
Willott, Elizabeth. 2004. "Restoring Nature, Without Mosquitoes?" *Restoration Ecology* 12 (2): 147–53.
Wylie, John. 2005. "A Single Day's Walking: Narrating Self and Landscape on the South West Coast Path." *Transactions of the Institute of British Geographers* 30 (2): 234–347.
Zelenski, John, Sara Warber, Jake M. Robinson, Alan C. Logan, and Susan L. Prescott. 2023. "Nature Connection: Providing a Pathway from Personal to Planetary Health." *Challenges* 14 (1): 16.
Zurawik, Marta. 2020. "Moving through Spaces—Leisure Walking and Its Psychosocial Benefits for Well-Being: A Narrative Review." *Human Movement* 21 (2): 1–8.

2

Placing Walking

Notes and Reflections from and About Kinetic Lifescapes

Alejandro Reig

The reflections recalled here are gathered from experiences that have taken place at different moments and in contrasting geographical and occupational contexts. They can be categorized as being at the intersection of auto-ethnography and methodological commentary. However, they are also self-help notes, useful for exploring our engagement with the world through and during walking. They are pictures in movement: initially reflections from the life of a migrant family and then the emplaced sketches of an independent adult migrant. Following these ethnographic reflections, I will explore how one model for understanding landscape processes can help approach these experiences and whether they can be interpreted through existing frameworks to address the embodiments of mobility.

I kick off this ball of reconstructed memories with a moment in which I can recover a memory of exploratory walking in the company of others: the long walks around my neighborhood in the east of Caracas in the late 1960s, when the city was still safe for children to wander around on their own. At least, these walks felt long to me then, as a child still in the single digits. Many of our excursions—to the supermarket a few blocks away, to the edge of our neighborhood imposed by the highway, or to the construction site about half a mile away—had something of a trespassing quality to them. We often walked to places where we were not supposed to go as children, sneaking into enclosed spaces at the limits of the urbanized areas, or doing any number of mischievous things—like stealing a can of anchovies from the supermarket to later eat in celebration. Apart from the contrast in safety and the scale of the places, which now seem much smaller, I highlight here the role of trespassing in the acquisition of walking independence.

I also recall some early walking on forest and scrubland trails through the Venezuelan wilderness, literally following in the steps of my father, an evolutionary biologist; one day, I carried a load of traps, the next, a handful of trapped mammals. I was warned to always keep an eye out for snakes along the edges of the path, a precaution that shaped one's attitudinal presence when walking the bush. This experience was as exciting as it was formative: you were given the responsibility to look out for yourself, on a path lined with dangers on both sides.

I now jump to a movement of heightened childhood autonomy: walking every day alone to school and back, at age nine with my six-year-old sibling, in West Hampstead, London, in the early 1970s. We would go to the public library, the park at Hampstead Heath, and on other urban explorations with my very multicultural "gang"—a group of seven, with just one British girl and one British boy. These walks also included trespassing: breaking into the basement of a church and snatching a brown beret from a box of old clothes for charity, which I would subsequently wear for many years, or jumping the fence into some neighbor's yard. We got shouted at by a lady who happened to be an arch-enemy of the gang: the headmistress of our school. One key excursion took us past Finchley Road tube station, up to Abbey Road studio; we vaguely knew that something important had happened at that zebra crossing. Many family walks with my parents also come to mind, including a memory of freezing our feet on Portobello Road, discovering Mao's Little Red Book at Speaker's Corner in Hyde Park—and getting my dad to buy it for me. Or a Black man standing on top of a crate, explaining that European culture came from Africa, in a direct line from the Egyptians to the Romans.

Our next family migration brought with it new, significant wilderness walks. Once again, I was a fledgling zoological fieldwork assistant, this time in southern Chile. I recall ankles challenged by the steep grassland slope of an Andean Valley along the Maule River, literally mined by the dens of underground rodents. Then a field trip to the Osorno volcano: on a slope of its temperate rainforest, I remember my first experience of walking in a wetland, leeches clasping themselves to my legs and slowly suckling. After drinking coffee made over a fire outside the tent in the morning mists, I recall a colleague of my dad's, a slim biologist from Wisconsin, shouting repeatedly at a sighting in a river torrent: "Oh gee, a puddle duck!"—my first impression of a scientist's violent exhilaration at discovering another wonder of the world.

The city of Valdivia, where we lived at the time, also allowed for exploratory roaming with friends around the wild environs of the university campus, on an island in the Calle-Calle River. We would trespass in the botanical gardens when it was closed, getting shouted at by Mrs. Taylor, a plump, energetic middle-aged Scottish woman who governed the fields and often came after knaves like us with a wooden cane. We would roam the nearby forest with handmade spears, reaching the riverbank and imagining we might be able to hunt some waterfowl—because fishing was boring and almost always unsuccessful.

And then from those years emerge a handful of formative observations of political walking. People marching and shouting in the streets: the workers and the students and the middle classes and the fascists—the last in neat squads. These memories are flanked by others just as strong: burning and burying books with my older brother in the backyard after the coup état, machine gun fire rattling at night while sleeping on a mattress on the floor, and my dad being released from prison and expelled from the country.

I recall just one significant episode of walking from our next migration stop, Argentina—both as a newly minted exile and a returnee. Again, accompanying my father, searching for fossils in the precipices and cliffs of Chapadmalal, on the seafront along the Atlantic coast. He moved slowly, making concentrated progress, minding his steps on the rocky and sandy ground, waves splashing nearby. His eyes followed the contours of the sedimentary structure of a thirty-meter wall, recognizing or "guesstimating" missing sections, intrusions of rock or organic matter, hunting for irregularities and emergent traces—sectioned humeri or femurs, joints, beaks, wings, or skeletal parts turned to stone. I followed attentively, scrutinizing both his gaze and the rocky wall, a proud son eager to learn to identify fossilized dinosaur, mammal, or batrachian remains—eager to participate, to be infused by a father's knowledge, and to try my own luck at discovering. Inflected in the present, this mode of walking I experienced then informs—I think—my own reflexive habit of discerning topographical structures and connecting with traces of the past, geological and archaeological, when wandering through natural landscapes.

I now jump to recall early teenage trekking in the subtropical Andes, specifically the Mérida mountain range in western Venezuela. First experiences with the weight of a backpack and deciding what to take on a journey, calculating times and distances, discussing weather, terrain conditions, and camping gear. The tension between letting go of things and the seduction of gadgets. Coming of age in the practices, discourse, and socialities of challenging walking, searching for the ethos of the frugal explorer. Getting hooked on this type of walking is about entering a set of experiential dimensions—exploratory, sportive, health related, but also transcendental (as in René Daumal's 2004 *The Analogous Mountain*)—that structure a certain aesthetics of living.

Urban walking also expanded during those years of residence in the Andes, with the freedom of lengthy walks with friends through a gentle city with a grid of streets and colonial architecture, nestled between two river valleys, and with nearly five-thousand-meter mountains facing the city. To this end, I pinpoint two specific situations:

1. Walking city streets, visiting shops or plazas or parks, but with eyes drifting toward the trails on the mountain cliffs and the silvery spots on the underside of the leaves of cecropia trees in the cloud forest. One walking practice becomes infused with the desire and the vision of another: the hike outdoors. The here and now is animated by an imagined translocation—a time-space project, as Torsten Hägerstrand (1982) might have put it—elsewhere.
2. Walking through topographical and architectural accidents, sometimes narrow or steep, rough sidewalks and city stairs, progressing by shifting one's gaze between the trail tread by the feet and the horizon ahead. The lyrics of a song, which I mention further on, will illustrate this.

During adolescence, my activities as field campaign assistant trapping mammals continued across different regions of Venezuela: coastal forests, mountains, and savannahs. These activities involved the same type of walking described earlier, but on a more regular basis, as they became a part-time job.

The next lifescape also unfolds in a city fronting a mountain. This time, it is not an Andean range but a high coastal range, El Ávila, which separates the modernist metropolis of Caracas from the Caribbean coast. Teenage life takes place in this city, structured by the erasure of its past and violent modernization, where opulent wealth and poverty are both obscenely evident. Chaos and beauty, an overabundance of vegetation, high-rise buildings and fast highways, with thousands of cars displacing pedestrians, rampant crime and police violence coexist in it—yet it is also filled with charm and the joy of living. In my city, walking can be dangerous and is, in fact, spatially and temporally delimited. But during the high school years, one begins to engage more with the city on foot, challenging no-go zones and no-go times. The tempo of the city is demanding but ultimately productive, with a sense of new possibilities always inebriating its inhabitants. The everyday contrast between the relentless rhythm and alertness of life and the colossal, looming presence of the Ávila mountain, a visual constant on the horizon, constitutes a defining existential feature of life in Caracas.

Then caving enters the picture. Trekking skills must be developed, training both swiftness and endurance, and walking is now done with a concrete aim: to discover underground worlds and map them into notebooks and papers for a speleological journal. The sociality of the whole endeavor is a key attraction of the activity: belonging to a new group, with a purpose, a culture, a history, and many stories to share, comprising a hierarchy of veterans and newcomers, heroes and commoners, with new disciplines and abilities to learn from. Learning to be this new type of trekker is a painful process, accomplished through exhaustion and physical and moral pain, enduring the cruel mockery of one's friends. Walking itself is never remarked upon. Conversations revolve around newfound caves, ropes, and ascent and descent techniques, fifty- or hundred-meter chasms, wild animals found while crawling along underground corridors, anecdotes of exploration, and histoplasmosis. But reaching a high caliber in walking is a must: to be accepted as part of the group, one must become a tough and reliable trekker who will not let one's partners down or slow the journey to or from the cave.

During my university years, two types of walking emerge. Moving around the UCV, the Central University of Venezuela, a vast microcosm of modernist tropical architecture and social, cultural, political, and educational diversity, holds a central place in the social fabric of those formative years. Intense intra-campus mobility, between faculties and social areas, becomes essential in this daily experience of discovery and initiation into adulthood. It's a sort of circular walking, moving between safe places to meet significant others; listen to lecturers, musicians, and political speakers; watch performances; and experience love and heartbreak. Outside the campus, these are years filled with walking in the city, day and night life, getting to know and live in new areas, challenging the no-go zones for partying, studying, and meeting people. But also, it marks the carving of the profile of an urban explorer of popular culture. Engaging with your feet in the city as a sociocultural landscape, brimming with other lives and stories to discover and experience—an ethnographer in the making.

Those years are marked by a decisive fall into the performative arts. Body and mind are abducted by a culture of rough, reclusive, and intense physical training, rooted in a formative tradition descending from Jerzy Grotowski and Eugenio Barba, and absorbing Gurdjieffian discipline. Years of searching for undivided presence, mastering will, and learning to pursue a project in frugality, against all odds. Also years of battering the body with acrobatics on concrete floors, while crafting time-and-space delimited poetic statements in the form of investigative theatre performances.

Yet this period of immersion into ascetic training and after-hours dedication eventually needs to come to an end, after many years of sacrificing normal life: regular income, quality time with a partner and friends during regular hours. A trek to the Andes mountains in solitude—amid peaks and glacier moraines, above four thousand meters in altitude—serves as this closure. A mountaineering friend sketches a plan for me of the way up to Páramo de los Conejos and the isolated Albarregas Lagoon. Each leg of the journey turns out to take twice as long to walk as he noted for me on paper, so I need to discard cargo at various points and make more sleeping stops than expected. Significantly, when I tell my friend about it after the trip, he agrees with my actual times: the competitive ethos of walkers typically includes some amount of fibbing about distances and efforts, perhaps to test and push one's potential walking mates or competitors. This trip becomes particularly memorable because I get lost along the path on my way downhill, then have to retrace my steps, adding three extra nights to my expected return time. My sister sends a warning to the National Park authorities, and I become the target of a search-and-rescue operation, complete with fliers featuring my photo. The entire trip turns out to be an ascetic encounter with my limits and with the gods who govern our path when we are in the mountains.

In the mountain facing Caracas, walking becomes a routine in three modalities:

1. The conversational walk with a partner along the firebreak pathway of the mountain, offering a scenic escape from the city.
2. Running the trails above the city: pushing one's limits to achieve aerobic enlightenment in the "second wind," subduing anxiety, and achieving something like a mercurial glide over the path while shaping one's life project and increasing awareness—running as mobile meditation.
3. Expeditions on the typical circuits of the coastal mountain range; falling in love with El Ávila (or "Waraira Repano") becomes a holistic exploration of what it is to be *caraqueño*, demanding periodic feats of discovery.

In all of these modes, the process of walking uphill from the city limits—twenty to thirty-five minutes of steep slope until reaching the landmark areas for rest or training—or along flat walking trails becomes a widespread social practice and a demanding physical exercise that one improves at over time. But it also provides a gradual, stress-relieving separation from the noise of the city and a deep dive into the sound, air, and vegetation of the green mountain. The *cerro* (or "hill," a rather modest name for what is, in fact, a larger mountain) affords city dwellers the resource of both a spiritual and physical routine, enacted in a variety of ways.

At a certain point, I read Peter Høeg's *Borderliners* (1994), and I realize that all that really matters in life is being the owner of one's own time. So I ask myself: what do I want to do for a living? I want to walk in the outdoors, meet people who are different, talk to them, listen to them, and write down their stories. I quit my job. Somewhere along the way, I am diagnosed with herniated lumbar discs. The doctor says I can no longer run. He advises me to strengthen my core muscles and avoid heavy weights. I switch to kayaking and cycling, but I continue trekking, lightweight: change is possible.

A significant shift takes place when I begin to walk in the Amazon rainforest. All that I had learned before proves useless. Lose the weight, walk with forest people, wade waist-deep through mud or water, keep your eyes on the trail for snakes, ascend and descend endless forested hills, endure the humid heat under the canopy, yearn for an opening and sighting of the sky—this, I realize, is what I always wanted.

I begin the journey toward becoming an anthropologist: building a network of relationships in the sciences and becoming a museographer of science, while simultaneously writing applications, researching scholarships, gathering reference letters, working, and waiting for my lucky break.

Later, my thesis fieldwork extends my trekking experience into the Yanomami mountains. I work to discern a pattern of being-in-the-forest-world, engaging with the environment through hunting and gathering. Gradually, I begin to register how foragers synchronize their activity with the lifecycles of the forest beings: the fructification of palms, the appearance and disappearance of fungi, bees, termites, other insects and their grubs, bird eggs. Their daily walks involve a multisensory following and assessment of traces and footprints—evidences of life and agency: chants, sounds, smells, footprints, animal tracks, their urine, their burrows; plants in which animals eat and nest, ecological associations (Reig 2013).

To follow the foragers, I learn to walk swiftly uphill and downhill, observing their technique, absorbing some of it. Scanning the foliage with one's eyes, adjusting the position of hips, knees, and spine to avoid fatigue. Eyes on the trail in front of one's feet, ahead, but also on the bough that could slash the face or eyes. Slippery trails and stumps on the ground pose a bigger threat than snakes. These people are shorter than I am; eventually, I need to squat and walk like they do or spend time clearing the overgrowth.

An interregnum of seclusion arrives, in the vaults of Oxford, to consolidate all that movement and reflection into a single thesis. In this *Alice in Wonderland* town of shrunken majestic buildings, transport mostly means cycling. Walks become rare outings to the nearby meadows, taken only when the writing monster allows a pause.

I pinpoint a long season of unsuccessful walking in Venezuela: years spent overtaking the streets with hundreds or thousands of people—not to get from one place to another but to occupy the public space with a demand for democracy. Useful in terms of galvanizing a collective demand and creating a space of identification in resistance to authoritarianism—but ultimately unsuccessful in reaching its goals.

The next move is settling into a Mediterranean landscape—an encounter with a new nature, thrifty in resources, without overabundance. Walking the Mallorcan mountains is walking a tamed nature trodden with history: our footprints are always, essentially, upon other footprints. Yes, but also . . . it is about brushing your ankles and limbs against thorny, small, woody plants, enjoying an epiphanic and nuanced light over the slopes, the not-so-loud murmurs of wildlife, discovering the silent sedimentary and metamorphic epics of geological becoming.

All of this brings to mind the irresistible resonance of a song from the 1980s: "The Fable of the Three Brothers" (Rodríguez 1980; translation by author). I paraphrase: three brothers set off on their life journeys "to discover and to build." The eldest went with his eyes fixed on the ground, watching his step to avoid going astray. But his neck got twisted for good—he became a slave to caution and grew old, still wanting to go far. The middle brother set out looking at the horizon. But this smart boy missed the rocks and holes in the path, fell again and again, and grew old, still wanting to go far, though he never made it. The youngest brother began his journey with one eye on the path ahead and the other fixed on his stride—one eye on the road and one on the future. But when he tried to sum up his journey, his gaze was already lost between being and going. The final lines go: "Listen to this and tell me what you think. An eye placed on everything knows not what it sees."

Everyday walking now takes place on the Via Verda, the Green Way, a twenty-nine-kilometer trail that begins close to my home and runs through four municipalities in eastern Mallorca. Developed along the remnants of an abandoned railway, it has become a hotspot for runners, hikers, and cyclists. Walking there has also served as part of my therapy for healing back and hip injuries. However, I find the flat rural landscape incredibly boring, so I walk trying to let go of the comparative prejudices shaped by the lush nature I grew up in. The trail is also threatened by those who want to rebuild the railway and remove the ecological pathway—a local expression of the friction between private, state, and public appropriations of the land, of citizen spaces either shrinking or resisting. But I fail to gather public interest in campaigning to defend it. This highlights my inability to connect with the local networks of environmental activism, which are largely defined by nationalism and antitourism sentiments.

However, I also try hard not to turn these outings into walks for thinking about complex and demanding issues, or for organizing my looming work schedule, but instead to actually enjoy simply walking in nature. I focus on the mindful experience of being in the walk, connecting with the sounds of the birds, goats, sheep, dogs, and hens, and the sight of rabbits and birds crossing the path. Achieving and maintaining this connection is not an easy task. But I realize it is also about processing the fate of the expatriate or exiled: letting go of the landscapes and lifescapes of your history that hold you back from being in the here and now, that keep you from embracing your present and its wonders.

A Stop in the Walk to Think: Kinetic Place-Making and Mobility Figures

I will now examine this collection of walking experiences through two distinct lenses: one that views walking as an embodied and (probably) universal practice and another that looks at its socially contextual categorization. The first approach offers a phenomenological exploration, asking: What kind of embodied experience occurs when we walk? The second approach analyzes the walks through various "figures" of mobility (Salazar 2017), which are understood as cultural tropes and historically constructed categories. While some features of these two sets may overlap, the exercise is good to think.

Interrogating the process of walking as an activity of "world-making," I am drawn to Eric Hirsch's (1995) model of landscape as a dynamic process, created through a back-and-forth movement between a background and a foreground. In this framework, a landscape is neither visual scenery displayed before an observer, the product of a historically constructed gaze that privileges certain ideas and perspectives, nor is it merely an environment shaped by natural and/or human forces. It is not a concretization of cultural-historical or natural determinants, a priori external to the subject. It cannot be abstracted from the agent's experience, as this landscape is always of *somebody*—an experiential configuration. The landscape belongs to a social actor who chooses, from a background of resources available within a certain (spatial-social-cultural-temporal) context, those that are meaningful to him or her, and actualizes them, bringing them to the foreground of lived experience. Through this process, place is made. These places become consolidated into landscapes through the actualization of their significance and their capacity to furnish our lived experience.

The usefulness of this processual view of landscape in approaching walking can be asserted in various ways. There is the parallelism between the background/foreground dynamics and the tension between the path on which we walk and the area surrounding it. A progression unfolds from the track to its environs, moving gradually from the distant to the adjacent, from the surrounding to the touchable, to a place where we stand. Walking can be seen as drawing a line with our movement across a medium composed of different planes, some of which we actualize by bringing them close to us, capturing them with our walk, while others remain untrampled, left at the sides or in the back, relegated. The spots we bring toward us comprise a minimal form of a place-making, produced through a succession of significant, valorized spots—"concretions of value," to echo geographer Yi-Fu Tuan's brilliant definition of place (Tuan 1977).

In this basic understanding of walking, places need not be symbolically important, pregnant with narratives and symbols, as they might be in cultural landscape or heritage approaches. In contrast to conceptualizations of landscape based on various forms of sacredness, I suggest that the ground captured by each step constitutes a place in the minimal sense: the materialization of a unique and unrepeatable spatio-temporal moment in our lives. However, we risk misusing this framework if we assume that every step taken is a valorized instant of experience (it holds value, yes, but is not necessarily valorized)—one that we would recognize as a place. Most steps are dots in a sequence of actions that do not become registered or embraced mindfully. Yet they are the building blocks of an experiential emplacement through which we construct our landscape on a medium scale, and that scale is the walk.

Establishing how walking constitutes place-making deserves further reflection, but I will hazard a temporary proposal. Place is made when an instant is affirmed—whether by slowing down or stopping to mindfully absorb a situation or location, or when a gestalt-like realization of a situation occurs in the momentum of continuous or discontinuous movement. Place is about realization, and this can take different forms when walking. As a disclaimer: in a broader sense, the track or path can also be considered a place, as can the area, neighborhood, region, and so on. However, this is a matter of scaling, metaphorization, and linguistic figuration.

A second, connected aspect of walking to interrogate when testing this place- and landscape-making framework is the issue of presence or attention. The ideal conception of walking for an explorer (or for a runner or other sportsperson) is one of fully resetting their attention in the experience, of consciously endorsing a "being-in-the-walk." From this perspective, we walk to explore a new and different space, becoming absorbed in the practice though a disconnection from our routine, and coming to discover new things in the world around us, whether in the proximity of our usual location or in more distant places. We expect and demand connection with these new places, offering them our full attention during our incursion. Clearly, this speaks from the biased preference of a certain kind of exploratory walking, and it is also an ideal of intermittent materialization. The expectation of being-in-place and in mindful connection is not something that can be permanently achieved during the experience of walking, and it does not need to be. Our attention may be on the new surroundings or within ourselves, far away, thinking about any aspect of our life that we need to sort out. Invaded by parasitic anxieties—about our job, social media debates, deadlines, the state of the world, financial hardship, etc.—we allow our pace to be captured by these thoughts, perhaps to macerate them and sort out issues though the momentum of the promenade.

We are not truly in the place in those moments but in our minds, our body transporting the whirling machinery of perpetual thought. Yet the walking experience allows for both disconnection and reconnection between these two modalities of attention. The abstracted-from-place inside and the emplaced-embodied connection with the outside world may come and go like a pendulum, with a potentially calming, digestive, or balancing effect on our spiritual and affective economies. And sometimes we might be walking with a partner, apparently wasting the chance to connect with incredible surroundings because we're busy talking. For instance, we might squander precious moments of attention to the winding way up from the monastery of Lluc in the Tramuntana range in Mallorca, across Mediterranean garrigue steppes and pine forests nestled among limestone massifs, to the Puig d'en Galileu (Galilee Peak), while catching up on accounts of job (in)stability or career progress with a friend and fellow walker-talker.

But setting aside this biased valorization of a certain type of walking over others, let's agree that walking is primarily a means of going from one place to another in already familiar settings. It's done without paying undue attention, just engaging in lightweight observation of our surroundings or our thoughts. In all of these cases, however, the dynamics of movement between the potential and the actual, the present and the absent, the materialized and the virtual, *take place*—in multiple senses of the term: they *happen* and they *capture a space*. In all these types of walks, we are moving between the perceived resources provided by the outside—whether material or ideational—and the actualization of some of these to achieve our micro-goals in our micro-projects: this is, to make places, valorized localizations that we may inhabit. In other words, when we walk, we perform a continuous and successive process of emplacement.

Summing up, we can say that walking implies a continuous process of place-making, where places are instantly affirmed and then abandoned as we move on to the next moment of place-making. This can occur both in mindful connection with the environment or while immersed in worries or imaginative wonderings—or in conversation or silent connection with someone else. Discarding preferences (for a walk connected with the outside world), let's acknowledge that place-making through walking happens in all three of these subjective engagements. In each of them, we sort out relevant issues or *realize things and places* by moving between possibilities and actualities, between the state of the problem and the options we can draw from the resources at our disposal, all of it channeled into the momentum of the repetitive pacing.

The walking vignettes developed in these pages serve as cases for this kinetic formulation of place-making I propose. They have occurred on a small scale, with events that could be replicated in similar or different situations, in other places, times, or by other people. At the same time, these stories are shaped by specific cultural, geographical, and historical forces. They are both unique and generalizable yet essentially socially inscribed. These are walks that take place in the life of a migrant, playing a significant role in the weaving of his lifescapes over time. How might they be viewed through categories of mobility, such as pilgrimage, pedestrianism, or flânerie, as discussed by Noel Salazar (2017)?

Some of these experiences could indeed be categorized as pilgrimage, if we define an explorer as a type of pilgrim—someone seeking connection with something that transcends everyday experience. This could apply, for instance, to a hiker in natural landscapes or an urban explorer venturing to previously untrampled places or time zones. In the urban explorations of my own early adulthood, I spot some semblance of flânerie, in rare moments when there was no specific destination, and one could be swept adrift by things, people, or signs appearing around the corner. However, usually there was some place to go to, an activity that structured the walk, so this improbable flâneur lacked the dandyism of idly hanging out, only to be swept up or away by the tides of city life.

During a period in my teens, I made it my project to fix broken bongos I discovered in the houses I visited. I embarked on a pilgrimage through music stores in rough areas of the city to buy goat-belly leather patches, then sought an ironsmith to solder the metal cylinders to the tuning rings. After weeks of effort, I would return, triumphant, to my girlfriend's family home with a fully functioning bongo. In the process, I discovered unimaginable corners of the city. This could be considered flânerie because desire was directing the prowess, but unlike the flâneur dominated by random happenstance, my exploration was organized within a purposeful micro-project of discovering city life.

In the spirit of being open to what the outside offers, could flânerie be found in the scientific fieldwork explorations presented earlier? Perhaps, but the idleness of the flâneur doesn't align with the focused self-conscious nature of research explorers. The early adult explorations in Caracas, which challenged the no-go times and no-go zones of the city, can be more accurately understood as the activities of a pedestrian, as described by Michel de Certeau (1984). These explorations reflect a desire to decode the clues and signs of the city, navigating its streets in unsuspected directions while seeking to incorporate more of it, going against the urban grammar and its dominant flows through speech-walk acts. Decades later, this defiant behavior was resignified when I ventured into impoverished areas of the city during times of political confrontation (counter to the habits of the educated middle class). This challenge to the limits imposed by fear worked by finding allies in places where confrontation could be expected. Today a courageous form of pedestrianism is carried out by urban activists in

Caracas, who work to reclaim city spaces in resistance to authoritarian power, as seen in projects like Ciudad Laboratorio (2024) and Labo Ciudadano (2024).

Many of my walks in the forest with the Yanomami were task oriented—expeditions aimed at reaching a village while supporting health care. Others constitute the "walking threads" of long-term research, seeking to understand "Yanomami place-making" and map "patterns of mobility," as I described in the thesis I later wrote and submitted for examination (Reig 2013). Pedestrian, flâneur, explorer? I feel these walks might be more aptly described as those of a cultural topographer—a term with a certain oxymoronic quality. Topographic work strives to objectively survey a positively existing material space through standardized scales and measurements; however, a search for cultural variation in spatial conceptions should also recognize different spatial ontologies. The cultural topographer may well be a figure of walking, but is there not always some element of pilgrimage in an anthropologist's fieldwork?

In different senses and at different scales, these experiences of walking could be used to inflect the categories theorized by scholars of mobility: pilgrim, flâneur, pedestrian. They are all encompassed here in a nomadic condition that extends from my childhood years of migration across continents to my current status as a member of a diaspora.

Coda and Closure

Bringing it all to temporary conclusive interrogations: In how many worlds and lives can we engage when we walk? We may actualize, as ritual does with myth, our formative walking perceptions and landscapes of life. We can experience our inner reflections and our perceptions of the outside. We can align them—or not. We may be welcoming the absolute outside, or we may be balancing the movement between our inner selves and the outside world, while also searching for our inside transformed by the reception of the outside. And for a migrant, all of this can become modified if we perceive our current situation as one of unfinished movement—a stage in a migration that allows aerial roots to graze and cling to new grounds, but ultimately with the hope (more than knowledge) that someday we will return. And, when walking, the Kavafian question arises once again: Can we create another new landscape and lifescape to call our own, or will we always perceive it through the shadows of our belonging?

References

Ciudad Laboratorio. 2024. Last accessed October 16. https://ciudlab.com.

Daumal, René. 2004. *Mount Analogue: A Novel.* Translated by Carol Cosman. Abrams.

de Certeau, Michel. 1984. *The Practice of Everyday Life*. Translated by Steven F. Rendall. University of California Press.

Hägerstrand, Torsten. 1982. "Diorama, Path and Project." *Tijdschrift voor economische en sociale geografie* 73 (6): 323–39.

Hirsch, Eric. 1995. "Introduction." In *Anthropology of Landscape: Perspectives on Place and Space*, edited by Eric Hirsch and Michael O'Hanlon. Oxford University Press.

Høeg, Peter. 1994. *Borderliners*. Translated by Barbara Haveland. Farrar, Straus and Giroux.

Labo Ciudadano. 2024. Last accessed October 16. https://www.instagram.com/labociudadano/channel.

Reig, Alejandro. 2013. "'When the Forest World Is Not Wide Enough We Open Up Many Clearings.' The Making of Landscape, Place and People Among the Shitari Yanomami of the Upper Ocamo Basin, Venezuela." DPhil thesis, University of Oxford, Institute of Social and Cultural Anthropology.

Rodríguez, Silvio. 1980. "Fábula de los Tres Hermanos." Track 5 on the album *Rabo de nube*. Spotify app.

Salazar, Noel B. 2017. "Key Figures of Mobility: An Introduction." *Social Anthropology/Anthropologie Sociale* 25 (1): 5–12.

Tuan, Yi-Fu. 1977. *Space and Place: The Perspective of Experience*. University of Minnesota Press.

3

Sentient Walking and the Good Life

Karen V. Lykke

A September evening in Oslo. Torrential rains have hammered down all day, and I am indoors, trying to capture what it is about the outdoors that brings solace to the soul, improves one's well-being, and makes life good. Venerated scholars and writers, from Thoreau (1862) to Solnit (2014) and Macfarlane (2012), have addressed these issues with such eloquence before, and trying to add another layer to their wisdom feels numbing. My dog is snoring beside me; every now and then, he sighs, signaling that it is here—in the dry, warm comfort of the living room, lying on a carpet in front of the fire—that he is most content at this moment. At some point tonight, nature will call, he will need to relieve himself, and I will be on the other end of the lead in the pouring rain. He likes walking, and once I get my boots and raincoat on, so do I.

There is something profoundly health inducing about these forced everyday encounters with nature while walking a dog—whether I walk on tarmac in the Oslo neighborhood where I live, wander the nearby woods whose periphery stretches into suburbia, or am fortunate enough to be by the sea, on an island in the Oslo fjord, traversing pine-clad paths that wind between rocks polished smooth by ice thousands of years ago. Walking wakes me up and forces me into awareness. Sometimes places in the landscape trigger memories or thoughts that have lain dormant until a sudden smell or sound or sight thrusts them to the very forefront of my consciousness. Seeing the trace or shadow of other beings outdoors gives me a profound sense of pleasure. Writing about these encounters is at once easy and challenging: Can anything novel or interesting be added to nature writing? And if so, what can I contribute?

Quite often, walking will trigger stories or memories that are not my own but rather collective ones—stories from the neighborhood, local history, folklore, ethnology, agriculture, and forestry, some of which are tied together by biological curiosities. As such, walking and sensing give rise to a mobile cabinet of curiosities of the mind—a Wunderkammer of interests and disciplines whose categorical boundaries are undefined (Impey and MacGregor 2001). When I walk, these curiosities accompany me, meandering along as they merge my conscious and unconscious thoughts, weaving together the past and present in both myself and my surroundings.

Michel de Certeau labels modern historiography as "heterography," or "a discourse about the other," because the past becomes an object of knowledge rather than an integrated part of history itself (de Certeau 1988, 3; Syse 2022, 99). While this perspective may be valid in many cases, I believe it is possible to make that "other"—whether a nonhuman other in the present or humans and nonhumans from the past—a close and integrated part of both the self and of historiography itself. Stepping back to an early modern worldview and curatorial practice, in which an assortment of precious, curious, and exotic objects, facts, fictions, mythologies, and sciences were mixed in a conglomerate under the direction of feudal and colonial power structures, is problematic for many reasons. Yet I believe that such an aggregation of thought also possesses healing properties—much like how dreams help cleanse the brain during sleep.

In this chapter, I attempt such an integrated historiography of agricultural land, woods, and suburbia, merging the self and the physical place of labor, chores, and movement, while also incorporating the senses in the process (Syse 2022, 102). As such, the personal healing action of sentient walking in the present is also a palimpsest of a multitude of actions, happenings, and social suffering in the past. As Jo Guldi and David Armitage argue, history has the ability to place people and events in a wider cultural context, particularly because it can claim to be a critical human science—combining the humanities and the social sciences to become "a tool of reform and a means of shaping alternative futures" (2014, 15). Sentient walking requires not only a certain embodiment of the landscape (Casey 1988, Ingold 1993) but also attention and imagination. Yi-Fu Tuan suggests that "a world imaginatively appropriated can seem more real than the one with which our senses are directly engaged" (1986, 160). I would like to push this idea further and build on Tuan by merging the sensed with the imaginatively appropriated, thus integrating both the past and the presently sensed in the self. These dense passages attempt to explain this framing in words; however, in what follows, I will largely let the words speak for themselves, allowing readers to reach their own conclusions.

My morning walk is always a short one, up what is more of a hillock than a hill, by Norwegian standards. Scramble to the top, and you are rewarded with a magnificent view of Oslo: a map-like panorama of old church towers, new high-rise buildings, parks, and roads—all framed in the distance by the Oslo fjord. On a clear day, you can see ferries, yachts, and the occasional cruise ship making their way to and from the harbor, while on overcast days, the city seems to shrink, pressed back and enveloped by the foggy sea air suffusing beneath an almost Turneresque sky.

The approach to this small hill is about two hundred meters on tarmac, to where my own road meets a long, sloping field, delineated by the metro station to the south and another road at its northerly end. Our local council keeps the field mowed—not with compulsive rigor, but with enough curative care to prevent the forest from encroaching on it. Right in the middle of the field stand a couple of huge trees: a massive Norway spruce, whose branches reach all the way down to the grass, and a triple-trunked goat willow with a rope ladder hanging from one of its thick branches. The trees provide shade to playing children on warm summer days and shelter to local pigeons hiding from dogs that chase them.

Once they gave shade to livestock. According to local history books, animals grazed on this land in the 1920s, before the farmer who owned it began selling it off, allowing suburbia to engulf what was once farmland and rough grazing areas. In the 1940s and 1950s, plots were sold to private housebuilders, and an amalgamation of houses sprang up, without much planning or aesthetic guidance from the local council. Although the area offers little in the way of grandeur or aesthetically pleasing architecture, it does have one sought-after asset: it's only a twenty-minute walk to the woods proper and, indeed, just a two-minute walk to a hundred-acre woodland that includes the aforementioned hillock. The name of the sloping field is Havnajordet, which literally means "the pasture field." *Havn* refers to a harbor, a fenced-off area, or a haven, and the word *have* or *hage* shares an etymology with the Anglo-Saxon word *hæge* and the English word "hedge" (Wright 2017, 44). An enclosed haven.

In winter, children sledge, ski, and even snowboard down the slope, pulling their gear back up the hill to go again. And again. In summer, teenagers lounge around, and the smell of burned sausages on disposable BBQs fills the air. Sometimes people sit reading on the four benches placed there—two at the top of the hill and two at the bottom—while the craze for fitness has worn a path down the middle of the field, as people doing interval training charge up and down. There are dog walkers like myself, and, in the early morning hours, people sprint down the hill to catch the metro, which runs every fifteen minutes.

It takes less than a minute to cross the field, take in the variety of action, and reach the edge of the woods. The path up to the little hill is steep and hardly visible from the cul-de-sac where I unleash the dog and let him run off into the woods. Tree roots crisscross the soil, clinging to it and keeping it from being washed down the hill in the rain. Sometimes small rocks and stones tumble down the path from the impact of boots and shoes and rain—my steps are an eroding force. Few people walk this path unless they have a dog—at least, I haven't encountered many.

The little woodland is diverse, though hazel is probably the most common tree here. I don't have much of a relationship with hazel—not like I do with the other trees in the woods, like the majestic Scots pine, which I almost feel I ought to greet as I pass, or the oak on the other side of the path that has a hole in its trunk where someone lives. My dog greets the hole every morning, pressing his nose into the cavity, snorting, checking in, while the oak stands proud, letting it all happen. An aspen, so tall I can't make out its crown as it stretches high above the hazel, also has a very prominent trunk and demands a kind of attention. I mistook it for a pine tree because its trunk was so massive and straight, its bark so confusingly brown and crinkled rather than aspen-gray. When the sound of its leaves gave it away, I was astonished and abashed for having confused two such common trees—or perhaps my embarrassment stemmed more from my own laziness. I hadn't looked properly nor listened carefully; I had simply assumed, walking past it for years, without looking up. And then, there, high above the hazel woods, I heard the shivering sound of the aspen. While perhaps I'm guilty of endowing a tree with personhood, I still feel a bit taken aback whenever I pass it—guilty of mindlessness, a lack of sensibility. Hazel, on the other hand, feels less like a tree, less noticeable, more utilitarian, less venerable. Clusters of rods, which could have been made into shepherding sticks or fencing in another world and forgotten past, simply poke out of the ground, through layers of pungently earthy-smelling leaves. Is it justifiable to neglect a stick of hazelwood, I wonder?

By early September, the cobnuts have ripened and changed from pale green to brown—a hazel brown. Two weeks earlier, part of the path had been covered in small brown chips from a spruce cone. On a branch above me, a red squirrel was ripping the seeds out of a cone, dropping the remains as it chewed its way around and around like corn on the cob, finally discarding the core on the ground before dancing farther out on the branch and starting to dissect a new cone. I felt a sense of joy from this small encounter, which inspired me to name these torn off bits of spruce cone "squirrel confetti," making me smile.

On this particular morning in mid-September, the path was covered in so many layers of it that some of the magic was lost. After all, the pale brown confetti doesn't light up the dark path in the same way when it's everywhere; the path just assumes the less enchanting, uniform color of Norway spruce cones. Still, it pleases me to think that every fragment of this bouncy layer, which my feet sink into, was torn off by the front teeth of a red squirrel.

Walking in Recent History

Sometimes I count my steps. I can stride up the hill in about 140 steps. I've done it in 130, too, but then it was more like leaping. I close my mind to the outer world when I count. About ninety steps up the path, there's a hollow stump, cut roughly forty centimeters above the ground. Seeing it casts my thoughts back to when this hill was last without tree cover, revealing how this landscape, like most, is a palimpsest of human action—some all but forgotten, many still visible even on daily dog walks.

At the top of the hill sits a small square house, with the year 1940 inscribed on it. As I mentioned, the views from here cover the naval approach to the Oslo fjord, and locals named the hill Tyskertoppen—the "German hilltop"—during the Nazi occupation. In 1943, the Germans built an air defense position here after losing the Battle of Stalingrad. Norway, with its long coastline, was strategically important for many reasons, and so "Festung Norwegen" was implemented: hundreds of Nazi fortresses were constructed all along the coast. Tyskertoppen was one of four permanent antiaircraft batteries in Oslo. According to signage placed by the local history group, it had four cannons and a commando shelter. At the time, the entire area was fenced in with barbed wire, and soldiers patrolled the fort with machine guns. Where there weren't soldiers, there were mines. Several machine gun sections were set up for local defense. The fort was also equipped with sound monitoring and bugging systems, searchlights, and armed concrete bunkers that housed the communication headquarters. While neither the signage nor local history books mention the deforestation of the hill, all the remaining stumps, the similar age of most of the trees, and common sense are proof enough for me.

The path from Tyskertoppen continues another fifty meters, meeting a road with a zebra crossing that leads pedestrians to the upper part of Havnajordet. In the morning, this is where my walk ends; in the afternoon and evening, it's where my walk truly begins. This area is flat. Two football goals have been placed here, and the field is delineated by houses with hedged gardens to the east and a massive stone fence to the west. The stone fence is quite an unusual structure in this part of Norway—it's more commonly found on the country's west coast. Known as Svenskemuren—the "Swedish wall"—the fence marks the boundary between Nordberg and the outfields of Brekke's farms. About a meter wide, it's crumbling a bit, but most of it still stands at about a meter tall. An impressive structure that winds for about 2.5 kilometers into the forest itself, the wall is said to have been built between 1807 and 1814 by Swedish prisoners of war during the Napoleonic Wars (1803–1815).

The path crawls alongside the stone fence, mostly fairly close to it, though diverging by as much as ten meters in some places, for no apparent reason. Perhaps these parts of the path predate the stone fence, as "[p]aths are habits of a landscape. They are acts of consensual making" (Macfarlane 2012, 17). They may have been made "by hooves and feet, walking habitually, going about every day or seasonal labour, every day or seasonable grazing, transporting people and things from place to place" (Syse 2022, 99). Using the stone fence as a visual navigational aid and sinking your feet into the sometimes quite muddy adjacent path takes you farther up and into the woods.

The Force of Nature

The local cross-country skiing club had hung a small light trail in the woods, maybe about three kilometers long. Suspended from tree to tree, it's very low tech, like something put up by eager skiers in the 1980s. It was used whether or not there was snow for skiing, making it easy to jog or walk in the woods during the afternoon and evening both in autumn and early spring, when the days are short. Dangling lights made from old-fashioned lightbulbs with aluminum shades directed the light down toward the ground rather than up into the sky. Naturally, they also marked the trail, whether or not the lights were on. The string of lights wasn't very invasive—certainly less invasive than if each light had been mounted atop a mast. In summer, when the days here are much longer than the nights, they were never switched on. Yet people still followed the trail in mid-summer—maybe because it's wider than the smaller paths, because it feels safe to follow a string of lightbulbs, or maybe just out of habit.

In early June 2024, a forty-nine-year-old man was found dead on the light trail. He had been out walking his two dogs, Zorro and Zimba, who were found sitting calmly—perhaps in shock—next to their lifeless owner. He was probably killed by lightning, though this wasn't immediately obvious. It took some time before the coroner and police aligned their conclusions. However, the evidence includes the fact that, close to where the man was found, eighteen trees lining the trail had vertical cracks splitting their bark along their trunks. In the local paper, a meteorologist explained how an electric field between a cloud and the ground builds up over time. At some point, a channel of charged particles descends from the cloud. As this channel nears the ground, the voltage becomes so high that a second channel forms from the highest point on the ground upward. When these channels meet, it generates extreme heat and causes an explosion—this is what we hear as thunder and see as lightning. The electricity then branches out over the ground, striking any trees, animals, or people in the area. A light trail like this, however, allows the current to spread more widely via the connected trees and wires, discharging the entire power channel from the thundercloud all at once. The electricity creates many branching currents in a relatively large radius from where the lightning first strikes, rather than just a single burst. Because the wires in the light trail were designed to conduct electricity, it's possible the lightning struck the wires and was directed straight toward the unfortunate man's body, or perhaps the lightning traveled through the trees, into the ground, and surged through his legs as he walked past (Buggeland and Hoftun 2024).

The local skiing club dismantled the light trail the following August. Only the eighteen trees with ripped bark remain, silent witnesses to a tragic death. They look as though their wood had suddenly grown too big for their bark casing. Maybe the story of the man will be forgotten, maybe not. In our digital age, local historians have an easier time uncovering incidents, accidents, and occurrences. Whenever I walk among the trees, I think of the man and his dogs, the immense power of the thunder and lightning, and how devastated his wife, two kids, and dogs must have been to lose him so abruptly. It's a story I tell anyone I'm walking with, a matter I discuss with people both in the know and not. It's not a story to be repeated too often outside the woods, but certainly one to reflect on whenever I walk in his footsteps and within the power circuit he found himself caught in. As Edward Casey (1988) wrote, "local history is literally a history of locales."

The even, fairly wide path that meanders through the open pine forest splits into three smaller ones as the terrain grows steeper. One path climbs a hill, another cuts straight alongside it, and a third runs downward. It's no more than a foot wide, with rocks, mud, twigs, and branches on it, and pine trees growing on either side that look like they're between forty and fifty years old. Or maybe it's the other way round—the path finds its way between the trees. My sense of logic corrects my intuitive prioritization of the path. It's unlikely that the path was there first and the trees then planted themselves on either side of it.

My footing isn't as confident as it was ten years ago—a knee injury has made me slightly less certain, and I grab the trees on either side of the path to steady myself. Pine bark feels rough and warm, and the trunks are easy to grip as I follow my dog's quick trot down the hill with less ease. "Four legs good, two legs bad," stated George Orwell's character Snowball, condensing the Commandments of Animalism in the face of human oppression (Orwell 1945). But Orwell probably didn't have the physical superiority of four legs in mind when he wrote it. Then a tree trunk breaks my rhythm—it feels polished and smooth, stripped of its bark. Dead. Rot, bugs, weather, and wind have stripped it. What killed it? Its top is splintered. Lightning? I walk on, mulling over the sheer planetary force of lightning and what happened in these woods earlier this summer.

In *Mrs, Dalloway*, Virginia Woolf tracked the thoughts of her protagonist using the stream of consciousness method, in an attempt "to depict the multitudinous thoughts and feelings which pass through the mind" (Cuddon 1984, 660–61). Woolf's narrative blurred and alternated between infinite awareness, understanding, indirect interior monologue, and simply having her protagonist talk to herself (Woolf, 1925). I talk to myself, with myself, and with my dog, in my thoughts, by muttering, and out loud. My own mobile cabinet of curiosities or stream of *un*consciousness talks and walks with me.

I think September is particularly interesting from a biological perspective. It's the meeting point between life, decay, and death, with all the fungi breaking down what were once living trees—photosynthesizing leaves—into soil, feeding off them, and creating something new. I like the way the roots crisscross the paths, how bright orange chanterelles sometimes protrude through this entanglement of roots, spruce needles, twigs, and dead leaves in the middle of the path, where hooves and feet have stomped, hardening the soil and presumably making life difficult—at least for green plants. Orange delicacies forcing their way through rot and decay. The roots look as though they are woven into one another, creating a supportive mesh of themselves and each other. Trees hold onto, grow into, and cover anything made by us humans.

Fighting the Elements

Along the path, another relic of the past is up for interpretation. At first glance, it looks like part of the aforementioned stone wall, but it's more like the remains of a stone igloo—one without a roof. The side that runs parallel to the path is only about a foot or two long, but one of its walls is at least a meter and a half high. It is built with smallish, flat stones, each less than a third the size of the stones in the wall itself. There are plenty of larger stones nearby, which might have been sturdier to build with but heavier to lift. Maybe the person who built the structure wasn't particularly strong? It would have been a very cold and drafty little hut, but its roof would have provided some cover from the snow and rain, and some of the inner stones appear to be the remains of a hearth, which would have warmed the humble hut somewhat. I wonder again: Did a weak or frail body build it? An old person, before social security and health care for all became the norm? Or a very young person? A pregnant girl, cast out of society? Maybe it had a dual purpose—perhaps it was also used to stable a horse back when every tree in these woods was turned into timber or firewood once it reached its peak. As late as the 1950s, the woods were still full of the sounds of barking spades and shouts of "timber" and would have smelled of sweaty horses pulling heavy logs. Another fifty years earlier, the sound of axes would have been just as prominent, as would the sounds of cows lowing and sheep bleating. There would have been children, laboring alongside their mothers as they tended to flocks of sheep and cows, or working alone. In Norway, in the late 1800s and even in the early 1900s, poverty among rural landless peasants was severe, and the archives are full of accounts of barefoot children with chilblains, starved and skinny, in both urban and rural settings.

On top of the hill, next to the stone fence, there are two heaps of sticks and stones. They are *varp*, or *kast*—places of remembrance. Technically, they are little piles of sticks and stones that have been placed to commemorate a harrowing event, like a murder or an execution. It's considered bad luck to pass such a place without paying one's respect by adding a stone or twig to the pile. The municipality has put up a sign here, written in seventeenth-century Norwegian: "*Udi Christian 4 dis Tiid dræbte een quinde her sit Barn. Hun bleff senere hængt i een Galge, og de tvende Hobe her ere deris Graffue*" ("In the time of Christian the Fourth, a woman killed her child. She was later hung in the gallows, and the two heaps here are their graves"). The ground here is rocky and shallow, virtually impossible to dig into, so if the incident did in fact occur, the heaps indicate commemoration rather than actual graves.

Local historians have also taken note of another story—of two Swedish prisoners of war who got into a fight and killed each other. It's possible that the *varp* commemorate this incident rather than an infanticide followed by an execution, but the first story is the official one now, and people continue to add to the heaps. An unmarked, narrower, but highly visible path runs off to the south of the two heaps. The woods here are open and light, with straight, tall Scots pines poking out of the reindeer lichen and other moss. It's a beautiful part of the woods, one of my favorite paths to follow. Less than 150 meters along this path, I found a photo and a letter in a plastic bag next to a tree. Another place of commemoration—a modern one. A man went into the woods and killed himself here. The first year, someone regularly visited the photo. More messages in the plastic bag; I managed to suppress my curiosity and didn't read them, but what else could the pieces of paper be? During the year that passed, the photo and words displayed in the plastic bag had begun to fade and dissolve. On one of my dog walks, I saw that the bag had been caught in the wind and misplaced. I put it back by its tree and placed a stone over it to keep it there. The next time I passed, the stone had been joined by three others, and now the commemorative plastic bag and letter have become a little pile of twigs and stones. Whether this tiny pile will grow into a proper heap one day remains to be seen.

I hear swearing, loud grunts, and shouting, and I look up. A man in his late teens or early twenties is riding an off-road fat bike, grunting and cursing on a narrow footpath that probably won't survive too many studded fat-bike tires before being shredded and eroded. Elsewhere in the woods, the traces of too many tires have already torn and widened paths, turning them into muddy highways. The young man's grunting and swearing may be a sign of an internal struggle against the hill, roots, stones, and his own fatigue—he is competing against either himself or some demon I can't see or hear. I step aside and curse them both under my breath, taking on the role of an evil sorceress myself. I'm astonished by the rage that builds inside me. I'm not sure whether he even saw me; it didn't seem like it.

I'm obviously not the only one noticing the magic of the forest and contemplating the layers of its history—so many books have been written about this. Yet this demon-chasing, forceful young cyclist made me contemplate my own awareness and the methodology of close reading the landscape.

Absorbing Through Walking

The cultural historian Ane Ohrvik describes in depth the method and methodology of close reading texts, explaining that reading not only requires the physical action of eye movement and perhaps a finger following the words, sentences, and paragraphs but also involves how the mind absorbs what is communicated at different levels of attention. Moreover, reading—and the meaning deduced from what is read—varies depending on the preconceived ideas and knowledge of each reader. She describes "the fundamental physical and cognitive aspects of reading as a distinct form of analytic attentiveness" (2024, 238). For me, as a cultural historian specializing in landscape and agricultural history, Ohrvik's approach adds to and echoes much of what I have both read and experienced in landscape studies.

An important contributor to this field is Christopher Tilley, who introduced a postprocessual phenomenological approach to archaeology in *A Phenomenology of Landscape* (1997). He explains how the embodied act of walking in the landscape and the concomitant intuitive interpretation of material culture are methodologically valuable for understanding the past. Feeling the land and approaching sites in the landscape using the same embodied movement as people in the past layers meaning, deepens understanding, and triggers emotive recognition. Some trees in the forest have been axed down, while others—like the hazel and the aspen—have spread their roots, colonizing wider areas. What archaeologists call "the structural context" may have changed significantly. What was once agricultural land may no longer be farmed. A Nazi fortress is no longer a military zone but simply a suburban hillock. Each layer of history within the woods can be experienced through the body and closely read—some with ease, others requiring more attentive consideration.

Yet the hill I walk is probably as steep today as it was one, ten, or fifty generations ago, and all movement will be another layer of the same advance. Paths may have grown wider but maintain the apparent continuity of presence and use. Stone fences, piled sticks, dividing my attention between seeing, listening, smelling, and staying upright—moving as my feet trail the footsteps and the paths—I cocreate and maintain a material movement heritage (Syse 2022, 98). Close reading this landscape, tracing and embodying this heritage, is an aesthetic and fulfilling act of remembrance and absorption: an unbidden embrace of the past, a recognition of how these woods will endure long after my own passing. Sentient awareness through walking encapsulates the good life for me. According to Yi-Fu Tuan, "Seeing beauty in nature or in an artwork may well produce a sense of physical well-being, but it is also a forgetfulness of self and a grateful awareness of the pure existence of another" (1986, 157). I tie the awareness and unconscious together, let it erupt in my everyday movement, in my everyday landscape, and keep on walking.

References

Buggeland, Sven Arne, and Guro Hoftun. 2024. "Døde trolig av lynnedslag—nå må idrettslaget rive lysløypa." *VG*, August 14. https://www.vg.no/nyheter/i/lwKQWk/doede-trolig-av-lynnedslag-naa-maa-idrettslaget-rive-lysloeypa.

Casey, Edward S. 1988. *Remembering: A Phenomenological Study*. Indiana University Press.

Cuddon, John Anthony. 1984. *A Dictionary of Literary Terms*. Penguin.

de Certeau, Michel. 1988. *The Writing of History*. New York: Columbia.

Guldi, Jo, and David Armitage. 2014. *The History Manifesto*. Cambridge University Press.

Impey, Oliver, and Arthur MacGregor. 2001. *The Origins of Museums: The Cabinet of Curiosities in Sixteenth- and Seventeenth-Century Europe*. Oxford University Press.

Ingold, Tim. 1993. "The Temporality of the Landscape." *World Archaeology* 25 (2): 137–284. https://doi.org/10.1080/00438243.1993.9980235.

Macfarlane, Robert. 2012. *The Old Ways: A Journey on Foot*. Hamish Hamilton.

Ohrvik, Ane. 2024. "What Is Close Reading? An Exploration of a Methodology." *Rethinking History* 28 (2): 238–60. https://doi.org/10.1080/13642529.2024.2345001.

Orwell, George. 1945. *Animal Farm*. Secker & Warburg.

Solnit, Rebecca. 2014. *Wanderlust: A History of Walking*. Granta Books.

Syse, Karen Lykke. 2022. "Hefting the Land: A Locative Heritage of Hooves and Feet." In *Pathways: Exploring the Routes of a Movement Heritage*, edited by Daniel Svensson, Katarina Saltzman, and Sverker Sörlin. White Horse Press.

Thoreau, Henry David. 1862. "Walking." In *Excursions*. Ticknor and Fields. Retrieved November 15, 2024, via Google Books.

Tilley, Christopher. 1997. *A Phenomenology of Landscape: Places, Paths and Monuments*. Berg Publishers.

Tuan, Yi-Fu. 1986. *The Good Life*. University of Wisconsin Press.

Woolf, Virginia. 1925. *Mrs. Dalloway*. Hogarth Press.

Wright, John. 2017. *A Natural History of the Hedgerow and Ditches, Dykes and Dry Stone Walls*. Profile Books.

4

Oceanic Mobilities

Walking-as-Method in Aquatic Environments

Jesse D. Peterson

Walking into the Sea

Walking into the sea becomes a struggle of wills. The silt, sand, or muck displaces, depresses, and deforms beneath the soles of the feet. Water tugs and swirls around the ankles. Waves push and pull against the thighs. Descending the decline, the body begs to become horizontal. The lungs ache for air, and the eyes will themselves to remain open. With the head submerged, water pressure increasing, the ground seems to slip away as each step struggles to find its grip—the body untethered now, wrapped in aqueous ether. Here, as with most things, the sea seems to dissolve the walk, suspend it, and incorporate the walker and their rhythms into itself.

Walking into the sea signals fundamental challenges for walking-as-method to accrue "knowledge" related to "unwalkable" spaces and times. Such "unwalkable" places include incredibly dense or extreme environments that require forms of mobility beyond walking—such as the oceanic depths, lake bottoms, outer space, quicksand, and vertical surfaces. These places are similar to but distinct from ones where walking is possible but forbidden or heavily policed, fenced, or controlled—places where walking is illegal, unsafe, prohibited, difficult, or off-limits. Considering walking in these unwalkable places highlights the context-specific horizontality and terrestrial and atmospheric biases of walking as a methodological approach. Indeed, walking is (typically) a land- and air-based affair. It must involve substances to walk through as well as a "ground" or surface to walk upon. It requires resistance to propel a walker forward and receptiveness to allow for walking direction and momentum. Consequently, when thinking about unwalkable places, we can more readily appreciate that walking also depends upon volumetric space—and, more specifically, on an environmental context with various material configurations that demand different strategies for navigating them.

Because the sea is a volumetric matrix layered with land, water, and air, it serves as a heuristic tool for thinking about and practicing walking methods. It has the capacity to make bipedal walking—an activity most humans learn around the age of one—less of a given. In other words, the sea offers potential sites for embodied methodological innovation in "walking," providing opportunities for mapping the contours of both the possibilities and limitations of walking-as-method. As a result, I am interested in exploring how aquatic spaces might defamiliarize walking methods, rendering them "strange," and how both practical and speculative incursions into these spaces can help test the limits of walking-as-method.

I was given the opportunity to reflect on this question in depth in 2022 when I was invited to participate in the One by Walking research network and to develop a series of workshops. Building on these experiences, this chapter argues that the multiple and shifting mobilities of the human body in space are just as well suited for walking in the sea as on land. It makes inroads into this claim by describing and reflecting on three workshops held in Norway, Finland, and Iceland. These workshops explored the fundamental challenges of performing walking methodologies that address aquatic challenges and other "inaccessible" spaces. The first workshop speculated on and innovated new ways of walking with the oceans, while the second trialed some of these methods. The third workshop sought to situate these methods within a wide spectrum of alternative mobilities and the more-than-human. Together the workshops aimed to attune us to the possibilities the sea offers for walking methodologies and to help us relate differently to aquatic spaces in general. To further explore the benefits of doing so, this chapter provides a methodological description of these workshops, which highlight concerns about who walks, how, and where (see Springgay and Truman 2018), as well as issues of transgressions and unfair arbitrations (Mavhunga 2014, 33) and experiences of the margins (Hadfield-Hill 2019, 172). Ultimately, the chapter seeks to explore how best to bring these places to life and share something significant about them.

Immersing the "Walk"

Walking with waters, or in watery spaces, is less strange than it might sound. The urge to return to sea (Shubin 2009)—to know it thoroughly and intimately—is not uncommon for us as "terrestrial" organisms, brought forth from the watery depths. As scholars continue to emphasize "the materiality of the waters that [we] are" (Neimanis 2017, 240), we can begin to adopt alternative perspectives that view walking as intimately connected to and with water—whether through the waters within us and our need for hydration, which are crucial aspects of identity and mobility (De Wolff et al. 2021; Hunter 2021), or through our hydrological impact on the landscapes we traverse, the "waterscapes" that sustain certain spatial relations fundamental to society (Karpouzoglou and Vij 2017), and our existence on a "blue planet" (Lotspeich 1998).

Moreover, as blue humanities scholar Steve Mentz argues, oceans have long functioned as sites where culture, technology, and the body work together to bind humans to "the bitter taste of saltwater [that] represents the real, alluring, and painful truth of the world beyond the human" (2020, 62–63, 102). Hence, by paying attention to water as embodied objects (such as seas and oceans), we can also explore differences with waters. In particular, swimming (Shefer and Bozalek 2022), ship mobilities (Anim-Addo et al. 2017), and wayfinding—a way of moving through space, including across seas, that relies on environmental knowledge and attunement (Fernandez-Velasco & Spiers 2024)—have all functioned as key practices that blend humans with the seas and can complement and extend analytical mobility lenses on walking (Salazar 2017).

Although seemingly straightforward, highlighting that human bodies are not uniquely adapted to aquatic environments points to ways in which walking on land is also heavily mediated by similar factors. Indeed, noticing how human bodies behave in water is significant because it is easy to take walking for granted. Rather than thinking of walking as something done upon, on, or over, it might be better to consider walking as a three-dimensional activity in which different materials entirely envelop the walker, akin to a worm tunneling through soft loam or a fish gliding through the sea. In this view, walking is understood less as a terrestrial than an atmospheric activity, and humans and other "land-bound" critters might be more rightly considered as air-dwelling (though not air-borne) organisms. Walking relies as much on feet touching the ground and the characteristics of the terrain as it does on gravity and the atmospheric pressure that allows the body to move in this way. By thinking of walking in volumetric rather than topographical terms, oceans and seas illustrate how walking-as-method might be better understood as a specific subcategory of a more general practice of mobility-as-method, with walking being one peculiar form that enlivens embodied, experiential knowing.

In adopting such lenses, we can learn to "walk" among the many kinds of waters that exist on Earth. These perspectives have the potential to offer critical correctives to walking methods, thereby challenging assumptions about who walks, how, and where (see Springgay and Truman 2018).

Walking 1: Hardangerfjorden, Rosendal, Norway

My first excursion with One by Walking involved inviting walking methodologists to speculate about ways of "walking" the sea that might address socioenvironmental challenges facing aqueous environments. As the group gathered along the Hardangerfjorden outside Rosendal, Norway, I asked participants to plan a "walk" on its seabed that would test the limits of the imagination and explore the margins of walking, mobility, path-making, and representation. To do so, I invited participants to form groups and develop an itinerary for a walk in the fjord.

Each group came up with unique suggestions related to walking-as-method in aqueous space. These included various interventions, such as walking coastlines, walking with experts, tracking seafood and the many pathways it takes, using technology to assist walking (for example, boats and fish finders), mobilizing the self differently (for example, wading instead of walking), and becoming octopuses (figure 4.1).

These interventions point to different modalities of walking in at least two ways. For one thing, walking-as-method often entails the use of some kind of support. While walking can be done barefoot or in minimalist footwear, shoes and boots are common tools that aid walkers in getting around. "Walking" the sea offers similar opportunities for employing technological supports, and the workshop participants were not shy in suggesting both existing and speculative technologies. One group, for example, proposed using an "underwater eye," which they explained could help reveal aspects of the ocean through its novel mode of mediation. Other technologies that could assist in navigating the space were also mentioned, including fish finders, trackers, boats, and submersibles. This is not surprising, given that humans are not well adapted to surviving (deep) underwater without some form of assistance. However, it is noteworthy because these responses highlight how walking methods—regardless of the environment—always involve some form of technological mediation between the self and place. This observation underscores the need for not only more

Figure 4.1 Groups came up with different "itineraries" for walking the sea, such as building an "underwater eye," walking and engaging with those familiar with the space, and "becoming" an octopus. Source: Original scans by Jesse D. Peterson of three of the walking itineraries made at the workshop in Rosendal, 2022.

critical reflection but also innovative adoption of a wide array of potential tools and resources that can filter, expand, focus, translate, or transform the encounters that occur while walking in any environment (see Jørgensen 2016).

In addition, developing an itinerary for walking the sea prompted participants to underscore the significance of the embodied sociality that also saturates the experience. The walking "grounds" were not necessarily conceptualized as empty vessels but as spaces that could be "felt" and "observed" by walking alongside other beings, critters, and substances that inhabit, traverse, and make use of these spaces. Furthermore, confronting the challenge of walking in inaccessible terrain invites reflection and speculation on how bodies mobilize, and how we might interact with and relate to these spaces in ways that more closely resemble the perspective of an insider than that of an outsider. As one group suggested, walking the sea would require becoming an octopus (figure 4.1). This speculative suggestion involved not only magical transformation but also an ambiguous, symbolic pragmatism: "squeez[ing] through the smallest spaces" and "feel[ing] with all your tentacles" could carry important methodological implications, emphasizing how knowledge might be made less anthropocentric and/or rooted in alternative, more-than-human perspectives (Peterson 2024).

Walking 2: Skärgårdshavet, Korpoström, Finland

The interventions proposed in the first workshop became the impetus and focus for the second, which took place in a local sea center and its science lab. I asked participants to implement their "walking" itineraries in the sea however they wished. Because this group was different from the first, I had to let go of continuity and give them free rein to do as they pleased.

I provided the participants with two prompts or requirements to guide our discussion at the end of the workshop. The first prompt asked participants to use their senses to acquaint themselves with the sea. I invited them to go outside and engage with the sea directly, offering them the option to use the resources of the science lab, which included waders, buckets, nets, Secchi disks, microscopes, preserved specimens, and more. I let them decide which tools they would use, how they would implement them, and how they would mobilize their bodies in the process. I encouraged them to pay attention to their choices, so we could discuss them later.

The second prompt invited participants to creatively, poetically, and didactically attempt to express an essence of the sea in any form they chose—whether through writing, video, sound, dance, or pilgrimage—either individually or collectively. In this way, this workshop built on the earlier one by encouraging participants to explore available technologies that could assist them in exploring an embodied sociality. The purpose of the second prompt was to capture an impression of their experience, noting how it was surprising, enlightening, or insightful. In short, I asked participants to do their best to bring this place to life and share something they found significant about it.

Due to the wind chill and temperature (around ten degrees Celsius during the day), several participants chose to stay out of the water. However, the rest made their way toward the sea, bringing many of the tools available with them. Some devoted themselves to searching for microscopic organisms in the water, others went on a leisurely seaside stroll in the sunlight, while a few busied themselves with some of the resources in the science lab. One participant took a net and, instead of using it to catch something, placed it over their head and began feeling through the mesh, imaginatively repositioning themselves in an aqueous state. I found myself working with sieves and preparing slides for the microscopes to offer an alternative lens for sensing the sea and some of its tiny organisms that would otherwise remain hidden to us.

While the first workshop allowed participants to theorize about possible human-oceanic exchange, this workshop exposed the materiality of these connections. Those who took part (myself included) were able to use a variety of tools commonly employed in marine science to plumb the "depths" of the sea and our knowledge of it in this particular location, even if the waterbody remained somewhat inaccessible. Our engagement, though heavily mediated through our equipment, bodies, and preconceptions, was still deeply visceral, highlighting how much of scientific research is thoroughly embodied. It is all too easy to assume that walking provides an affective, embodied encounter that other methods may not account for. But walking the sea positions walking-as-method as unique in its modalities rather than its effects, especially in comparison to other field-based approaches. All participants—whether they engaged directly with the sea or remained in the lab exploring the specimen collections and books—were able to bring some of the sea into their respective approaches and find meaning in the process. Left to their own devices, participants enacted an attentive, intentional engagement with the sea that enriched both their experience of place and their understanding of the relations between the self and the sea.

Walking 3: Gudrúnarlaug, Dalabyggð, Iceland

Though held in relative proximity to the sea—because Iceland is an island and we were somewhat near the coast—the third workshop was located too far inland to access it. As a result, the participants were unable to work with deep waters and instead had to settle for walking in the vicinity of a local stream. Focusing on the theoretical insights from the previous workshop—particularly that the materiality of water redirects our mobility and attention—I asked the participants to explore walking as a multifaceted practice within situated mobile ecologies, where waterbodies, such as seas and oceans, represent a general type.

For this final workshop, I asked participants to pay attention to their individual experience of walking, to examine how it surveys not only the world but also the "terrains" of one's mind (Solnit 2014), as well as the intersecting pathways and arbitrations that accompany the act of walking. In this exercise, I directed the group to walk down the Sælingsdalsvegur, to the point where it crosses the stream by a bridge. Descending from the road, I asked participants

Figure 4.2 Participants moving in a single direction according to the gerunds presented to them in the workshop. Source: Jesse D. Peterson, 2023.

to arrange themselves in an outward-facing circle, shoulder to shoulder. Each participant was then given a mobility modifier—gerunds scrawled on slips of paper, such as "meandering," "skipping," "ambling," "scurrying," or "hiding". These modifiers emphasized the unfixed and shifting nature of walking (or mobility), highlighting the various moods and forms that the act of walking can take. The participants were then instructed to move in the direction they were facing, adjusting their walk according to the modifier they had received. Due to the location, they had to trudge through fields, ascend the local hillsides to the west, or walk along or across the stream itself. The intention was to allow the "lie of the land" to also "act" upon each walk, giving participants the opportunity to make choices about where to walk and how (figure 4.2).

Upon regrouping, workshop participants commented on how the modifiers not only influenced their manner of movement but also their experience of the place itself. One participant mentioned that, while walking, their focus changed from feelings of grief and concern to a form of celebration with the land. Another found their attention drifting from the ground to the sky, allowing the path they were creating to merge with the direction and flow of the white clouds above. A third participant plunged into the stream, trekking up and down the rocky channel.

Instructing participants to walk in a single direction and add character to their movements accentuated the challenge of how bodies engage with space and how spaces, in turn, engage with the body. Through this activity, walking could take on a particular feel or transmute into something akin to but not quite the same as walking—wandering, ambling, strolling, marching, drifting, bobbing, swimming, paddling, floating, and so on—additional mobilities the body can assume that imbue the movement and "walking" style with symbolic meaning or purpose. Moreover, even within the strictest parameters, walking does not operate the same way within different cultures or even different places. Walking is both an individual and situated practice, but it is also learned within and influenced by social parameters. It takes on a unique form and character in each situation. As geographer Kimberly Peters notes, "There is a need . . . to unpack the specificity of particular ways or methods of being mobile as each of these has its own cultural connotations, affective registers and political purposes" (2015, 8). Further, it is not only categories of mobility that require unpacking but each instance of mobility. For walking is never monolithic; rather, it is fragile, momentary, and unstable, responding to and modulating itself in every instant.

Attending to the "particular ways or methods of being mobile" opens the mind to consider not only the affective registers of one's own body but also those of other beings and other paths. Participants found it difficult to move in a straight line due to the stream, the undulating terrain, grasses snaking toward the sky, berms, clouds, buildings, fences, and roads. The world imposed itself as participants intersected with the "technological reinforcements [and] unfair arbitrations," raising the question of "who is transgressing on the road of the other" (Mavhunga 2014, 33). By occupying a different mentality, expressed through an alternative modality of mobility, participants came to understand the space and their relation to it in fascinating new ways (Salazar et al. 2017).

Back to the Surface

Having dabbled in the waters in the different locations of the three workshops, I argue that waterbodies (and other unwalkable places), in all their shapes and sizes, can reveal fundamental concerns about walking and what walking comes to mean. By attending to walking-as-method in the context of the sea, the workshops explored the limits of walking methods as terrestrially bound. The participants and themes that emerged from the workshops generated inquiry, reflection, and innovation on mobility, technology, intentionality/attention, and the ambiguity that attends every walk in any space. The physical act of walking can take on different tones or characters depending on the walker's intentions, their attention to place, the place itself, and the encounters that may, do, or do not occur. By attending to walking in aquatic environments, one can begin to unravel various assumptions and biases about walking while also noting fundamental and general features that can be applied to alternative forms of human and nonhuman mobility. Additionally, this practice fosters environmental awareness of places of tremendous importance not only to us humans but to other creatures as well. To use walking as a means of developing knowledge about ourselves and our place—as one among many—in the world points to the importance of thinking through how we walk, how our walk modulates, how it is assisted, and how it conforms to or disrupts a sense of embodied knowing. By walking into, around, and out of the sea, I suggest that walking-as-method might be a fruitful way to actualize attention to any space—including attention to the spatial and temporal means of doing so—no matter how "unwalkable" such spaces may seem.

As feet find a sloping bank, the head begins to emerge. Water splashes, drips, and runs. The sound of air rushes into the ear canal, harmonizing with a seagull's shriek. Light floods the eyes. Sunrays strike the shoulders as the torso dips, then lifts and expands once again. The wind blasts past, evaporating water, raising the skin. The body settles onto the buttresses of its knees and ankles. The feet continue to walk as the sea releases the walker onto sand and clay, one after the next, uninterrupted.

References

Anim-Addo, Anyaa, William Hasty, and Kimberly Peters. 2017. "The Mobilities of Ships and Shipped Mobilities." In *The Mobilities of Ships*, edited by Anyaa Anim-Addo, William Hasty, and Kimberly Peters. Routledge.

De Wolff, Kim, Rina Faletti, and Ignacio López-Calvo. 2021. *Hydrohumanities: Water Discourse and Environmental Futures*. University of California Press.

Fernandez-Velasco, Pablo, and Hugo J. Spiers. 2024. "Wayfinding Across Ocean and Tundra: What Traditional Cultures Teach Us About Navigation." *Trends in Cognitive Sciences* 28 (1): 56–71.

Hadfield-Hill, Sophie. 2019. "Margin." In *Keywords in Radical Geography: Antipode at 50*, edited by Antipode Editorial Collective, Tariq Jazeel, Andy Kent, et al. Wiley.

Hunter, Victoria. 2021. "Watery Bodies and Watery Sites: Immersion and Porosity." In *Site, Dance and Body: Movement, Materials and Corporeal Engagement*. Palgrave Macmillan.

Jørgensen, Finn Arne. 2016. "Walking with GPS: An Object Lesson." In *Methodological Challenges in Nature-Culture and Environmental History Research*, edited by Jocelyn Thorpe, Stephanie Rutherford, and L. Anders Sandberg. Routledge.

Karpouzoglou, Timos, and Sumit Vij. 2017. "Waterscape: A Perspective for Understanding the Contested Geography of Water." *Wiley Interdisciplinary Reviews: Water* 4 (3): e1210.

Lotspeich, Frederick B. 1998. "The Blue Planet: Mankind's Habitat." *Bulletin of the Ecological Society of America* 79 (3): 222–25.

Mavhunga, Clapperton Chakanetsa. 2014. "Seeing the National Park from Outside It: On an African Epistemology of Nature." *RCC Perspectives* (1): 53–60.

Mentz, Steve. 2020. *Ocean*. Bloomsbury Academic.

Neimanis, Astrida. 2017. *Bodies of Water: Posthuman Feminist Phenomenology*. Bloomsbury Academic.

Peters, Kimberley. 2015. "Drifting: Towards Mobilities at Sea." *Transactions of the Institute of British Geographers* 40 (2): 262–72.

Peterson, Jesse D. 2024. "Seeking an Algal Perspective: Exploring 'Harmful' Algae through an Interview with *Nodularia spumigena*." In *Being Algae: Transformations in Water, Plants*, edited by Yogi Hale Hendlin, Johanna Weggelaar, Natalis Derossi, and Sergio Mugnai. Brill Academic Publishers.

Salazar, Noel B. 2017. "Key Figures of Mobility: An Introduction." *Social Anthropology* 25 (1): 5–12.

Salazar, Noel B., Alice Elliot, and Roger Norum. 2017. "Studying Mobilities: Theoretical Notes and Methodological Queries." In *Methodologies of Mobility: Ethnography and Experiment* (Vol. 2), edited by Alice Elliot, Roger Norum, and Noel B. Salazar. Berghahn Books.

Shefer, Tamara, and Vivienne Bozalek. 2022. "Wild Swimming Methodologies for Decolonial Feminist Justice-to-Come Scholarship." *Feminist Review* 130 (1): 26–43.

Shubin, Neil. 2009. *Your Inner Fish: A Journey into the 3.5-Billion-Year History of the Human Body*. Vintage.

Solnit, Rebecca. 2014. *Wanderlust: A History of Walking*. Templar Publishing.

Springgay, Stephanie, and Sarah E. Truman. 2018. *Walking Methodologies in a More-than-Human World: WalkingLab*. Routledge.

5

Walking with Sensitivity

Emily Höckert, Outi Rantala, and Veera Kinnunen

Invitation to Stay Proximate

The planetary climate and biodiversity crisis compel us to critically reflect on the onto-epistemological assumptions that underpin humanist and social-scientific knowledge production. This requires the development of new conceptual and methodological approaches that address the interdependencies of the well-being of all living communities. Our research group Intra-Living in the Anthropocene (ILA) has focused on exploring proximity as an alternative, affirmative way of engaging with the ecological crisis (Rantala et al. 2024). Proximity, as we understand it, encompasses both scale—a mode of staying geographically close—and affective closeness—a way of being in relation. Researching with proximity unsettles the abstract narratives of the ecological emergency by exploring minimal, situated, and speculative ways of knowing from a "down-to-earth" perspective (Tsing et al. 2019; Zylinska 2014). It encourages engagement with mundane, local, and familiar sociomaterial processes, prompting reflection on the interdependencies and entanglements of more-than-human lives (Bell et al. 2017; Clark and Szerszynski 2020; Gibson et al. 2015; Laakso and Aro 2022).

Our exploration with proximity has been inspired by emerging feminist new materialist and multispecies discussions that challenge the prevailing, anthropocentric understanding of well-being (Alaimo 2010; van Dooren et al. 2016). These research streams aim to move the focus beyond human communities by exploring the inherent entanglements between more-than-human agencies—such as plants, microbes, fungi, soil, human and nonhuman animals, and water in its various forms. They seek to cultivate attentiveness to more-than-human relations and recognition of how other-than-humans produce and share knowledge (Bubandt et al. 2022; Rantala et al. 2024).

These discussions highlight material ways of being in the world, such as the interactions of human corporeality with the other-than-human world, and the processes through which knowledge comes to be (Alaimo 2010; Ulmer 2017). In our attempt to "proximatize" research—to put our thinking with proximity into action—we have been developing methods that seek to activate the situated, affective, and corporeal aspects of knowing. Proximate methodologies thus aim to grasp relational modes of being by developing new forms of sense-making.

In our previous work, we have explored a variety of embodied exercises that enable attentiveness to the experience of being with the other-than-human world. This has involved asking, for instance, how to sensitize oneself to the situated entanglements of bodies, elements, and atmospheres (Höckert et al. 2022; Kinnunen et al. 2021; Rantala et al. 2024; Salmela and Valtonen 2019). Our methodological approach to sensitizing is influenced by feminist new materialist scholar Sari Irni's (2016) method of "sensitive reading," which calls for a more fruitful and less categorical encounter between poststructuralist and new materialist scholarly discussions. To this end, we have engaged with Irni's method of paying special attention to nuances, power relations, and feelings in order to disrupt prevailing binaries between nature and culture, human and other-than-human.

Drawing inspiration from Irni's (2016) insights, we have developed the method further by transforming it from a textual into a corporeal exercise and taking the exercise outdoors. As an additional step, we have focused attention on human encounters with other-than-human companions, testing different ways of communicating the nuances, aspects of power, and feelings that arise in these encounters (see Höckert et al. 2022; Salmela and Valtonen 2019). What follows is an example of a "walking-with" experiment, in which the three dimensions of sensitivity introduced by Irni (2016) facilitate sensitization to interdependencies entangled in our relational way of being in the world—without romanticizing or reducing difference to sameness (Höckert et al. 2022; Viken et al. 2021). We call this experiment "walking with sensitivity."

According to Irni (2017), sensitive reading may enable "seeing-anew" by learning to "see-with." In our case, we suggest that sensitive walking-with other-than-human companions can help us rethink our relationality with "nature" and nonhuman animals. Expanding on Donna Haraway's (2016) concept of "becoming-with," walking-with conveys an ethical standpoint that deviates from anthropocentric norms and hierarchical ways of thinking and acting by referring to a considered, relational, respectful, or reciprocal approach between the walker and other-than-human companions (Cahoon 2025). Thus, we propose that walking with sensitivity, as a method, can enable us to engage with our proximate surroundings in more curious, caring, and speculative ways.

By testing and developing walking with sensitivity as an experimental method, we join the rich tradition of walking as a research methodology in the social sciences and environmental humanities (for example, Ingold and Vergunst 2008; Springgay and Truman 2018). The exercise described in detail here can serve as a playful tool for speculating about new ways of relating to other-than-human companions in diverse contexts, as well as a methodological guide for gathering empirical material during fieldwork and field courses.

Script for the Experiment

We tested and developed the walking with sensitivity exercise in slightly different forms in three One by Walking "walkshops," which were held in Rosendal, Norway (May 2022); Vålådalen, Sweden (September 2022); and Korpo, Finland (April 2023). In brief, the exercise consists of four steps: orienting to the exercise through instructions, walking with a freely chosen other-than-human companion (see figure 5.1), processing the experience through writing, and sharing and speculating about the experiences with other human companions. Participants are encouraged to arrive with a weather proof attitude and clothing, pen and paper, and a preparedness to be otherwise unprepared. We hope these hands-on—or, rather, feet-on—how-to instructions are seen as recipes that can

Figure 5.1 Walking with flowers. Source: Photo by Camilla Brudin Borg and Emily Höckert.

be freely adapted to suit different occasions, tastes, and needs. The vignettes following each step offer an example of one participant's experience from the May 2022 walkshop in Rosendal, Norway.

Script for the "Walking with Sensitivity" Experiment

To begin the exercise, put these three concepts into your metaphorical pocket or backpack: nuances, feelings, and power relations. Next, take an attentive, lingering, slow walk with an "other-than-human" companion. After the walk, process the walking experience through speculative writing. As a final step, share and develop your experience further. In what follows, we explain these four steps in more detail.

Step 1: Instructions

The purpose of this walking-with experiment is to cultivate sensitive modes of attunement toward other-than-human communities. To this end, we ask that you choose one other-than-human companion for "walking with." You can choose your companion once we start the walk—that is, whenever you feel ready to invite someone to join you. The companion can be material, such as paths, mosquitoes, pavements, or rocks, but it can also be spiritual or conceptual, like sounds, wind, or hidden beings. Whatever companion you choose, make sure that it is proximate in the sense that you can "feel" its presence, whether in a conceptual or material form.

Next, please place these three concepts into your metaphorical pocket or bag: nuances, power relations, and feelings. The walking experiment aims to explore how these concepts might orient you differently toward other-than-human companions that inhabit or are present in this place. This could mean perceiving and sensing the variety of nuances in your walking companion, reflecting on questions of power between different beings—such as humans and their chosen walking companions—and being attentive to any feelings that arise during the walk. At first glance, these concepts—nuances, power relations, and feelings—might seem abstract and vague, and each walker will likely understand them differently. However, we do not want to offer strict definitions, as these concepts are good for walking and thinking with precisely because of their openness.

We are gathered here in a circle next to a big, park-like area and grazing sheep. I'm curious about the small flowers with purple petals growing everywhere and would like to pay closer attention to them. I'm not sure whether I can concentrate on all those concepts—nuances, power relations, and feelings—simultaneously. Perhaps I will focus most on power relations, as I'm eager to question the predominant idea of perceiving nature as merely a resource for humans. Or maybe, because I have this strong urge to relate differently, I'm also bringing along feelings. . . . And while much of the communication here has involved speaking, being "alone" with these plants will give me a chance to test more nuanced ways of communicating. (Participant's reflection from the walkshop in Rosendal, May 2022)

Step 2: Walking with

Now we ask that you turn off all gadgets and walk around the area. Take your time to invite "your" other-than-human companion along with you. You have twenty minutes to walk silently, without speaking to human companions. Take time to stroll and choose your route intuitively, letting your walking companion guide you. Engage all your senses, paying attention to the subtle presence of your other-than-human companion.

Look around and begin to make observations by tuning into your feelings, sensing your body, and being attentive to even the tiniest details. Feel with your hands, feet, and face. Remember to notice smells, as the sense of smell is strongly connected with emotions and memory. As you walk, allow different kinds of memories to join you. Pay attention to what kinds of beings you and your companion are connected with. What kinds of worlds—on different scales—can you attend to? Carry the concepts of nuances, power relations, and feelings with you in the back of your mind. Let them guide your walk without becoming too analytical or critical.

I carefully enter the group of flowers and ask them to join me as my walking companions in this exercise. I promise to be a gentle companion, rather than one who crushes them with my shoes. Inhaling their scent, I realize that this is how I engaged with new flowers as a child. I pick up my phone automatically, take close-up pictures, and try to capture the panoramic view—before realizing that I'm going against the instructions. I feel guilty for wanting to capture this abundance and beauty rather than simply enjoying their presence. If I were closer to home, I would probably want to dig up a few plants and bring them to my garden. I want to make them mine, even though there might be similar flowers at home that I've just overlooked. This makes me wonder how the local people might see and relate to these plants. What if the flowers are considered weeds or invasive species to be controlled or eradicated? They may even be poisoned, like dandelions are in many places. Or maybe they have been used as medicine and celebrated in folklore and handicrafts. Funnily, I had initially thought I was the flowers' only guest or companion, but then I became aware of all this buzzing other-than-human life around, among, and on them. The other walkers are already gathering, but I wish I could stay longer. (Participant's reflection from the walkshop in Rosendal, May 2022)

Step 3: Processing through Writing

The time is up, and we come together again as a group, including the other-than-humans that were not as vividly present at the beginning of the exercise. Now we kindly ask you to find a comfortable place to sit in silence for a while and dwell on the impressions and observations you made while walking with your chosen companion. When you are ready, take your pen and paper and jot down your thoughts, questions, sensations, and affective curiosities. Keep writing without criticizing, thinking, hesitating, or controlling your process. You can also draw. You do not have to show this paper to anyone unless you want to.

Continue reflecting on the nuances, power relations, and feelings you noticed while walking. You may use the following questions to guide your writing:

- Who did you choose to walk with? Why?
- What kind of experience was it to walk with this companion?
- Did you pay attention to atmospheres, things, or beings you haven't noticed before?
- Did you feel that something was overpowering other elements, observations, or beings?
- Did anything arouse strong feelings?
- Did anything spark your curiosity or raise new questions?
- What kinds of questions are you posing now?
- Who would you like to invite to explore your new ideas further (for example, beings, disciplines, experts, arts, species)?

This exercise offered a chance to visit, slow down, walk, and stay with local flowers. Being with them made me aware of my somewhat, and lamentably, automatized thinking patterns that revolve around capturing, owning, and focusing on myself. I wanted to approach the flowers, wondering what color their petals were; but when thinking about nuances, it became less important to give a color an exact name. That is, to categorize and claim to know the other. It's possible to engage with these flowers without knowing their Norwegian, English, Finnish, or Latin names. But considering the world-making power of naming, it would be naïve to think that the names humans give them don't matter—whether they are called "Vomit Flowers" or "Angel's Sighs." It's pretty clear that I'm not a natural scientist here, and being a stranger in this place leaves me with many questions I'd like to explore further.

I remember my colleague Ana María mentioned that Nancy—or was it Lacan?—refers to humans as "the speaking subjects," which sets us apart from other-than-humans, from nonspeaking subjects. Although verbal communication is only a tiny part of all more-than-human communication, I catch myself prioritizing it as a primary form of relating with other humans. But after just a short visit among the flowers, I long for more of this other kind of bodily communication that occurs through the senses and in silence—but not in the absence of sounds or other messages. (Participant's reflection from the walkshop in Rosendal, May 2022)

Step 4: Sharing

As the last step of this exercise, please share your experiences with the other participants. While you are welcome to share any kind of experience, including difficult ones, you are also free to keep some thoughts to yourself. Please share with us whether the experiment "did" anything to your thinking or to your relationship with the beings inhabiting this place. Did the exercise prompt any new questions, hesitations, or revelations?

When using this exercise as methodological guidance for gathering empirical material with more-than-human communities, you can also reflect on whether it provided any new insights into your ongoing research project(s). Propose ideas for what might be done next: How could this exercise guide the fieldwork you are doing or about to begin?

I tell the others that I walked with the mesmerizing field of flowers over there and learned that the plant is called the "Cuckoo Flower" or "Mayflower." It grows in many places in the Northern Hemisphere, is edible, and has also been used as medicine to treat various diseases. I say a few words about the multiple ways and nuances in how "we" humans might have related to different flowers over time. For instance, growing up in a modern Western society, I learned only very limited ways to relate to flowers: knowing their names, picking them for aesthetic purposes, or even pulling them up with their roots if I feel like it. To claim to "know the other" just because I know their name is a form of epistemic violence that can lead to other forms of misuse of power.

Thinking about feelings entangled with nuances and power relations, I feel embarrassed to confess that, for a brief moment, I thought I was the flowers' main guest or companion. However, they are hosting and caring for many forms of life

> *in this meadow and beyond. From here on, I'd like to continue exploring how these "Cuckoo Flowers" have been presented in local folklore and handicrafts. In my following research walks, I want to keep seeking ways to bridge the prevailing nature-culture division that separates humans from other-than-human communities.* (Participant's reflection from the walkshop in Rosendal, May 2022)

Wondering with Other-than-Human Companions

For us, the three aspects of sensitivity—nuances, power relations, and feelings—offer an alternative to the Western, Anthropocenic way of making sense of the world, which is based on binaries between self and other, humans and other-than-humans, nature and culture, subject and object. Together these concepts can help us recognize how proximity can intensify relations and should be treated as an ongoing, uncertain process of becoming, rather than a desirable goal. Our approach to sensitivity thus refers to accepting the open character of being and knowing in proximate relations (see Zylinska 2014). It is a mode of attending to relations that eludes closed endings while acknowledging the multiplicity, difference, and uneven distribution of more-than-human responsibilities (see also Haraway 2016; Hylland Eriksen et al. 2018; Neimanis et al. 2015; Pálsson and Swanson 2016; Todd 2016; Tsing 2015).

More-than-human methodologies encourage us to hesitate, speculate, and wonder, so that multiple voices can be recognized and heard. In line with Harawayian thinking, our walking-with experiment was designed to foster a curious and caring orientation toward our proximate surroundings, beings, and thoughts. We consider enhanced modes of wonder and speculation the main takeaways from this exercise and believe that researchers can use the experiences from the walks in a variety of ways.

It should be noted that, even though scholars like us are often obsessed with producing data—and even though our perception is often steered by an instrumental and productive mindset—this exercise is not fundamentally about gathering data or bringing back trophies from our walks. Instead it is about making space for the possibility to think differently. For us, the exercise is about creating room to ask new kinds of questions. Thus, rather than providing research data, this exercise might help shake up preconceived and learned ways of perceiving the environment, opening up curiosity and wonder, as well as new questions and research avenues.

Sensitizing ourselves to different modalities of engagement can enrich the perception of a landscape as storied and shared with other-than-human companions. While the example of the exercise described here accentuates feelings, power relations, and nuances, using different concepts could direct attention to other phenomena. For instance, concepts such as care, vulnerability, and reparation might encourage walkers to focus on relationality and perhaps more affirmative modes of more-than-human relations (Rantala et al. 2020). Regardless of the concepts walkers bring along, the exercise is designed to prompt attention to human-centeredness and raise questions such as: How can we attend to more-than-human modes of being and be attentive to their stories? How can we disrupt our human-centered experience of being in the world?

Acknowledgments

We would like to express our gratitude to all the members of the ILA group, the One by Walking network, and everyone who has walked with us. Special and heartfelt thanks go to John Martin and Dominica Williamson for inviting us to further develop the methodology through the wonderful concept of "Tiny Landscape Stories."

References

Alaimo, Stacy. 2010. *Bodily Natures: Science, Environment, and the Material Self.* Indiana University Press.

Bell, Sarah, Lesley Instone, and Kathleen Mee. 2017. "Engaged Witnessing: Researching with the More-than-Human." *Area* 50 (1): 136–44. https://doi.org/10.1111/area.12346.

Bubandt, Nils, Astrid Oberborbeck Andersen, and Rachel Cypher, eds. 2022. *Rubber Boots Methods for the Anthropocene: Doing Fieldwork in Multispecies Worlds.* University of Minnesota Press.

Cahoon, Neal. 2025. "Researching 'with' Through Practice-Research Methodologies." *Sosiologia* 62(1): 20–38. https://journal.fi/sosiologia/article/view/159735/103596.

Clark, Nigel, and Bronsilaw Szerszynski. 2020. *Planetary Social Thought: The Anthropocene Challenge to the Social Sciences.* Polity.

Gibson, Katherine, Deborah Bird Rose, and Ruth Fincher, eds. 2015. *Manifesto for Living in the Anthropocene.* Punctum books.

Haraway, Donna. 2016. *Staying with the Trouble*. Duke University Press.
Höckert, Emily, Outi Rantala, and Gunnar Thór Jóhannesson. 2022. "Sensitive Communication with Proximate Messmates." *Tourism, Culture, and Communication* 22 (2): 181–92. https://doi.org/10.3727/109830421X16296375579624.
Hylland Eriksen, Thomas, Sanna Valkonen, and Jarno Valkonen, eds. 2018. *Knowing from the Indigenous North: Sámi Approaches to History, Politics and Belonging*. Routledge.
Ingold, Tim, and Jo Lee Vergunst. 2008. *Ways of Walking: Ethnography and Practice on Foot*. Ashgate Publishing.
Irni, Sari. 2016. "Kun Jälkistrukturalismi Kohtaa Luonnontieteistä Inspiroituneen Uusmaterialismin: Herkän Luennan Harjoitus." *Naistutkimus* 4: 5–16.
Irni, Sari. 2017. "Fluid Reading Practice: On the Queer Potential of Studying Nonhuman Animals." *Feminist Encounters: A Journal of Critical Studies in Culture and Politics* 1 (1): 05. https://doi.org/10.20897/femenc.201705.
Kinnunen, Veera, Sandra Wallenius-Korkalo, and Pälvi Rantala. 2021. "Transformative Events: Feminist Experiments in Writing Differently." *Gender, Work & Organization* 28: 656–71. https://doi.org/10.1111/gwao.12606.
Laakso, Senja, and Riikka Aro. 2022. *Planeetan Kokoinen Arki—Askelia Kestävämpään Politiikkaan*. Gaudeamus.
Neimanis, Astrida, Cecilia Åsberg, and Johan Hedrén. 2015. "Four Problems, Four Directions for Environmental Humanities: Toward Critical Posthumanities for the Anthropocene." *Ethics and the Environment* 20 (1): 67–97. https://doi.org/10.2979/ethicsenviro.20.1.67.
Pálsson, Gísli, and Heather Anne Swanson. 2016. "Down to Earth: Geosocialities and Geopolitics." *Environmental Humanities* 8 (2): 149–71. https://doi.org/10.1215/22011919-3664202.
Rantala, Outi, Veera Kinnunen, and Emily Höckert, eds. 2024. *Researching with Proximity: Relational Methodologies for the Anthropocene*. Palgrave MacMillan.
Rantala, Outi, Anu Valtonen, and Tarja Salmela. 2020. "Walking with Rocks—with Care." In *Ethics and Politics of Space for the Anthropocene*, edited by Anu Valtonen, Outi Rantala, and Paolo D. Farah. Edward Elgar.
Salmela, Tarja, and Anu Valtonen. 2019. "Towards Collective Ways of Knowing in the Anthropocene: Walking-with Multiple Others." *Matkailututkimus* 15 (2): 18–32. https://doi.org/10.33351/mt.88267.
Springgay, Stephanie, and Sarah E. Truman. 2018. *Walking Methodologies in a More-than-Human World: Walking-Lab*. Routledge.
Todd, Zoe. 2016. "An Indigenous Feminist's Take on the Ontological Turn: 'Ontology' Is Just Another Word for Colonialism." *Journal of Historical Sociology* 29 (1): 4–22. https://doi.org/10.1111/johs.12124.
Tsing, Anna Lowenhaupt. 2015. *The Mushroom at the End of the World: On the Possibility of Life in Capitalist Ruins*. Princeton University Press.

Tsing, Anna Lowenhaupt, Andrew S. Mathews, and Nils. Bubandt. 2019. “Patchy Anthropocene: Landscape Structure, Multispecies History and the Retooling of Anthropology.” *Current Anthropology* 60 (S20): 186–97. https://doi.org/10.1086/703391.

Ulmer, Jasmine B. 2017. “Posthumanism as Research Methodology: Inquiry in the Anthropocene.” *International Journal of Qualitative Studies in Education* 30 (9): 832–48. https://doi.org/10.1080/09518398.2017.1336806.

van Dooren, Thom, Eben Kirksey, and Ursula Münster. 2016. “Multispecies Studies: Cultivating Arts of Attentiveness.” *Environmental Humanities* 8 (1): 1–23. https://doi.org/10.1215/22011919-3527695.

Viken, Arvid, Emily Höckert, and Bryan S. R. Grimwood. 2021. “Cultural Sensitivity: Engaging Difference in Tourism.” *Annals of Tourism Research* 89: 103223. https://doi.org/10.1016/j.annals.2021.103223.

Zylinska, Joanna. 2014. *Minimal Ethics for the Anthropocene*. Open Humanities Press.

6

Possibilities in Motion

Prosthetic Maps and Micro-Walks for Tectonic Encounters

Alice Lewis

Everything is in motion in and around us all the time. From the microscopic cells that combine to form our familiar human selves to the great pulses of shifting planetary orbits, everything is constantly on the move. Around 4.5 billion years ago, specific elements came together in particular astrochemical relations, leading to the formation of Earth and, over time, the emergence of living organisms—including humans (Grosz 2008, 6). This reconfiguration of relations continues to occur at scales and over time frames so vast that they are largely incomprehensible to us, yet they continually shape our lives on Earth. Contrary to what humans might like to think, the Earth is far from stable; it is in a state of constant mobility, reorganizing human and more-than-human relationships—and our own lives—in the process. Recognizing and inhabiting the reciprocity between these incomprehensible geologic motions and our human actions and patterns of inhabitation is essential for unsettling tropes of human exceptionalism and understanding our role as participants in the vibrant life of this planet.

The *Everything Is Moving, All the Time* project explores how we might encounter the reciprocal relationality of geomorphic events and human patterns of inhabiting Earth. It was carried out through a series of site-specific interventions in Australia, Canada, and Finland between 2022 and 2023. In each intervention, participant groups used purpose-designed prosthetic maps and micro-walks to inhabit discursive sites of relation between individual, personal human temporalities, and the ongoing geologic processes in each location. This chapter details the project and its approach through one specific participatory intervention that took place in the Finnish archipelago, in the country's southwest, with the One by Walking research group. The maps and walks paralleled participants' life events with the geologic processes of Glacial Isostatic Adjustment (GIA), which has been occurring in the region since the end of the last ice age, around ten thousand years ago. The larger project and the tools and techniques developed for it aim to help participants better comprehend the vast planetary motions and shifts happening around us all the time, with the intention of finding new ways and possibilities to inhabit Earth in a post-Anthropocene era.

Everything Is Moving

Viewed from the globally dominant Eurocentric scientific and intellectualist perspective, the Earth is a surface upon which human life plays out (Ingold 2010; Tsing 2015). However, the recent popularization of "the Anthropocene"—the idea that humans have altered Earth's ecosystems to such an extent as to warrant the recognition of a new geologic era—demands a rethinking of the Earth as an inert environment. As geographer Kathryn Yusoff points out, by recognizing humans as geomorphic agents acting on the ground, we must also acknowledge our geologic relationality, because our inhabiting of Earth's terrestrial environment is a reciprocal situation made possible by Earth's geologic motion (2015, 6). Processes of geologic formation play an active role in our own geographical and social formations—including the development of land, materials, and fossil fuels—and, thus, in determining what it means to be human. Consequently, while we humans may be recognized as geomorphic agents shaping the Earth in the twenty-first century, the Earth's tectonic processes also continue to shape us, reducing humanity to just one actor in the participatory, relational milieu that forms Earth.

Occasionally, the active world-building agency of nonhuman geomorphic events becomes visible through earthquakes or volcanic eruptions, which prompt mass human migration and new configurations of inhabitation. For the most part, however, geologic processes and their reconfiguration of human social and geographic relations occur over timescales and through forces so vast and gradual that they are difficult to comprehend, and easy to overlook. In describing similar processes, cultural theorist Elizabeth Grosz (2008) uses the phrase "cosmological imponderables" to refer to "the forces of temporality, gravity, magnetism . . . [which] are among the invisible, unheard, imperceptible forces" that operate outside the control of life, even as they create the very conditions for life to not only exist but continue to evolve (23). These processes unfold in forms, orders, specialities, and timescales that exist simultaneously with our own yet largely remain outside the reach of modern scientific imagination, making them difficult to grasp. Nonetheless, traces and signs of this ongoing reciprocity between life and geologic motion exist all around us.

The "geography of geomorphic events" and the possibilities afforded for human inhabitation become visible when plotting European cave art created around the time of the Last Glacial Maximum, twenty-six thousand years ago, when extensive ice sheets covered the planet. The locations of European cave art from this period show a pattern of inhabitation that followed the edge of the snowline (Yusoff 2015, 394). While the ice sheet that covered much of Europe at that time has long since receded, the geomorphic event of GIA, set in motion by the formation and subsequent melting of ice, is still ongoing today. The weight of ice, which gathered in sheets up to four kilometers thick across the Northern Hemisphere, pressed the Earth's crust into the liquid aesthenosphere below. As the ice melted around ten thousand years ago, the pressure was relieved, and the crust began to rebound, slowly moving back up into its original position. This process of GIA is still occurring today across northern regions of both North America and Europe. In the Finnish archipelago, this process of rebound causes the Earth's crust to rise out of the Baltic Sea at a rate of six millimeters per year, resulting in new terrestrial land—called "uplift land"—and the subsequent development of the Finnish Land Uplift Act to regulate ownership rights, resource management, public access, zoning, and development in this new terrestrial environment. The geomorphic events of glaciation and GIA have been shaping human geographies, laws, and society for millennia and continue to do so today.

Encountering and recognizing the reciprocal relationality of geomorphic events and our human inhabitation of the Earth helps to unsettle ideas of human exceptionalism. Reconsidering human hierarchies in more-than-human environments is crucial as we come to understand both our dependence on more-than-human entities for survival and our largely unabated destruction of them. In finding ways to encounter and understand humanity's role within a milieu of actors, however, Yusoff (2015) cautions that flattening the human/more-than-human hierarchy entirely would negate the responsibly we have inherited as humans (400–401). Human responsibility arises simultaneously from the legacy of our destruction and the capacity for repair afforded to us by our humanity. Attempting to rise to this challenge, the walking maps developed for *Everything Is Moving, All of the Time* aim to highlight our individual roles as active yet relational participants in these world-building processes. Facilitating this type of encounter with Earth has the potential to open new possibilities for how we might inhabit the planet, both now and in the future (Yusoff 2015; Springgay and Truman 2018; Grosz 2005).

Prosthetic Maps for Geologic Encounters

The prosthetic maps developed for the *Everything Is Moving, All the Time* project aim to draw together site-specific geologic temporalities and human experience. Prostheses are human-made technologies that we add to our bodies to facilitate alternate modes of inhabiting Earth. While in medical terms "prostheses" refers to the replacement of perceived embodied deficiencies (Brill 2011, 245), the use of the term in this project draws on Grosz's (2005) assertion that the continual reformation of our human capacities, through enmeshment with prostheses, opens up new performative possibilities and ways of engaging with the world (152). The tendency of humans to augment our bodily capabilities by interlinking with prosthetic technologies has allowed us to enact significant spatial, social, and political changes on Earth, making us "geomorphic agents" and helping to develop the Anthropocentric condition we now inhabit. However, this same power afforded to humans through our capacity to augment our bodily abilities to enact change means that we must now assume responsibility for the present state of the Earth and redirect our energy to care for that which supports us. To do so, we need to find new ways of seeing, knowing, and acting.

Maps are always, already prostheses. Generally, they are Cartesian cartographic visualizations that offer both the creator and the user a top-down, or "god-like" perspective of the world (Specht and Feigenbaum 2018, 40). Today maps are ubiquitous, embedded in our smartphones and GPS devices. While many of us take these visualizations to be accurate representations of places, they are, in fact, powerful prosthetic devices for forming and reconfiguring the world and how we can (and do) act within it. As geographer Denis Wood (2010) points out, maps do not simply represent the world as it is but are "instruments for its creation . . . systems of propositions, arguments about what the world might be" (8). In other words, the authors of maps exert power by determining how we understand, and thus inhabit, Earth. This political and cultural agency has been—and continues to be—employed by colonizing powers to suppress human communities and assert ownership over "discovered" places (Specht and Feigenbaum 2018, 40). Increasingly, however, this same prosthetic power of maps to reconfigure narratives of place is being subverted to reveal previously disregarded realities, reconnect subjugated peoples with their lands and languages, and counter capitalist logics that obscure the detrimental impacts of resource-heavy modern life on landscapes.

In the *Everything Is Moving, All the Time* project, the prosthetic power of maps and tools of Cartesian cartographic visualizations are used to illustrate the "imponderable" (to borrow Grosz's term) geologic temporalities that shape specific landscapes. The purpose-made maps employ a range of representational forms, including timelines, map projections, and diagrams, all drawn onto single printed broadsheets. Folded like traditional hiking maps, they serve as tools for carrying out interventions that explore specific landscapes. Participants use the maps as guides to conduct short walks across land that has been geologically altered during their lifetime, thus simultaneously inhabiting both human and geologic temporalities.

The maps perform two key functions within the interventions. The first is to equip the reader with the necessary information to understand the pulsating rhythms to which modern science attributes the geologic change occurring in the intervention locations. The map developed for Korpoström, the island in the Finnish archipelago where the intervention detailed in this chapter was held, highlights cycles of glaciation (ice ages) and resulting GIA occurring across vast spaces and time periods. As seen in figure 6.1, a timeline tracks Earth's identified glacial periods over the last four hundred million years, and a set of diagrams depicting the whole Earth illustrates dominant scientific beliefs about why periodic cooling of the planet's surface temperature occurs. The hypothesized causes include changes to Earth's elliptical orbit (from elongated to more spherical every forty-one thousand years) and the changing tilt of Earth's axis over one-hundred-thousand-year periods—both of which are believed to contribute significantly to cooling the planet's surface temperatures due to proximity to the sun—as well as continental drift, in which moving landmasses alter ocean currents and the distribution of heat through water. These movements occur over imponderable timescales yet are believed to have contributed to a drop in Earth's surface temperature, which led to the massing of ice over the Northern Hemisphere during the last ice age, impacting human inhabitation patterns

Figure 6.1 Map developed for Korpoström illustrating geologic process of glaciation, with a focus on Finland. Source: Photo by Alice Lewis, 2023.

up to the present day. A collection of drawings also tracks the basic process of change from ice melt at a global scale to the zone of GIA, which is causing Finland and adjacent countries to rise vertically out of the Baltic Sea at varying rates. However, these drawings are still subject to the limitations inherent in the top-down cartographic perspective, particularly in the abstraction of the place in question—both in conveying spaces so vast they are difficult to relate to the self and in omitting details simply due to scale.

In negotiating the limitations of the map itself, its second key function is to act as a tool for fostering encounters between the user, their own life events, and the visible geologic changes occurring at the intervention sites. In the Finnish archipelago, geologic processes can be seen in the ancient boat-mooring rings attached to rocks that now sit meters away from the water's edge; however, the rate of this change remains abstract. To move beyond abstraction, a ruler on the right-hand edge of the map sheet marked six-millimeter rises—the rate of GIA in one-year increments at Korpoström—providing a way to relate the ongoing geomorphic event to human years. This measure forms the basis for self-directed micro-walks and creates an important point of convergence between pulsating geologic movements, the timescales of a human map user's life events, and the actualities of GIA on the ground. Rather than offering suggestions for specific ways of inhabiting the archipelago long term, the practice of linking specific local movements with nonlocal planetary motions that are always occurring around us aims to render visible the active and ongoing agency of geomorphic events in both shaping our social formations and, as Yusoff (2013) notes, determining what it is to be human.

Micro-Walks for Traversing Timescales

Walking as a means of connecting to environments is a well-established practice (Springgay and Truman 2018, 3). Anthropologist Tim Ingold (2010) describes walking as a form of "ambulatory knowing" (S122), whereby the walker accrues knowledge of environments by moving through them and paying attention. This process of movement along a path outlines an exchange whereby the walker encounters the environment from various (moving) perspectives while also leaving imprints on it. Though the impacts on both may be small, the process always involves an exchange. The *Everything Is Moving, All the Time* project facilitated participants in the process of ambulatory knowing by guiding them through individual micro-walks specific to their own life events. During the intervention at Korpoström, the micro-walking paths traversed the short distance between the sea edge and the terrestrial "uplift land" that has emerged from the sea over a perceptible time period related to participants' own life experiences, thus connecting localized experience with geologic forces.

The intervention began with a collective reading of the maps with all participants to highlight the processes of motion occurring around us at that very moment. Walking together to the water's edge, participants used the embedded map ruler and a range of other ad hoc measuring devices to plot their own micro-walking paths across uplift land that has emerged over the course of a specific, personal life event. They then set off, slowly treading their individual paths. These micro-walking paths may have been short—often only a meter or so in length—but in walking this limited geographical expanse, the map user began to inhabit the discursive space of exchange between their own human experiences of the terrestrial world and the processes of geologic movement illustrated on the map.

While the physical space in which the body moved on these micro-walks was relatively small, the actual "sites" of the event were not the physical place of Korpoström. Rather, they arose from the friction that occurred at the intersection of the walker's own human time and geologic time. The human-geologic timescales drawn into relation created discursive sites of exchange between the walkers and the vast temporalities and spaces of geologic movements. Discursive sites are spaces of relational exchange between entities (human or more-than-human) where meaning is made and realities—new, altered, or persisting—emerge. By measuring the emergence of uplift land against one's own known time periods, these two temporalities are brought into relation, offering a means to comprehend the "imponderable" forces and magnetisms that generate gradual but ceaseless movement around us all the time. Therefore the "site" being "walked" was not merely the path identified by the walker on the ground but the space between our human activities and the geomorphic processes that stretch backward and forward through eons on their slow trajectories. Encountering these parallel timescales has the potential to reorganize both human and more-than-human entities by revealing them as entities that participate, in their own ways, in our inhabitation of Earth.

Toward Future Possibilities

Finding ways to inhabit the many timescales of movement that shape Earth and our inhabitation of it—including the sociality and politics of being human—is crucial for uncovering the possibilities afforded by recognizing that everything is always in motion. Doing so "might open into a more generous ecological thought" (Yusoff 2014, 402). Thinking generously about ecology and the more-than-human temporalities that make human life possible is becoming increasingly important as we begin to comprehend the damage being done to our living environments by our persisting ideas of Earth as an inert surface with resources for human consumption. Leveraging the power of prosthetic technologies, such as maps, to counteract these pervasive perceptions and develop new relational narratives is central to sustaining life on Earth. The techniques of Cartesian cartography borrowed for this project can offer humans an overview (god like perspective) of spaces and places not accessible otherwise, but the abstraction and distance inherent in mapmaking give the author power to decide what is drawn and left out—and, by extension, to shape how people actually inhabit those places in practice. Consequently, these tools must be used carefully and creatively to mitigate their inherent limitations.

The ongoing *Everything Is Moving, All of the Time* project seeks to counter the abstraction inherent in the prosthetic maps developed for each intervention site by integrating micro-walking events into their use. The walks, plotted against the walker's own life events, are carried out in mapped sites to allow participants to fill in any gaps in the drawings through lived experience and register the parallel timescales of planetary motion and human inhabitation of Earth through their own bodies. This embodied, active practice helps to negate the limitations and abstraction of Cartesian cartography by enabling the inhabitation of both timescales—that of our familiar selves and the incomprehensible scale of geologic time—simultaneously, on the map and on the ground. It also facilitates walking the discursive sites of exchange that arise in the friction between them. Rather than dictating a potential end, this project, carried out through a series of interventions, develops tools and approaches for understanding the world in motion, opening up possibilities that arise from understanding these geologic forces for future ways of being and inhabiting Earth.

References

Brill, Sarah. 2011. "The Prosthetic Cosmos: Elizabeth Grosz's Ecology of the Future." *Philosophy Today* 55 (Supplement): 245–54. https://doi.org/10.5840/philtoday201155 Supplement29.

Grosz, Elizabeth. 2005 *Time Travels: Feminism, Nature, Power*. Duke University Press.

Grosz, Elizabeth. 2008. *Chaos, Territory, Art : Deleuze and the Framing of the Earth*. Columbia University Press.

Ingold, Tim. 2010. "Footprints Through the Weather-World: Walking, Breathing, Knowing." *Journal of the Royal Anthropological Institute* 16 (s1): S121–S139. https://doi.org/10.1111/j.1467-9655.2010.01613.x.

Specht, Dough, and Anna Feigenbaum. 2018. "From the Cartographic Gaze to Contestatory Cartographies." In *Mapping and Politics in the Digital Age*, edited by Pol Bargués-Pedreny, David Chandler, and Elena Simon. Routledge.

Springgay, Stephanie, and Sarah E. Truman. 2018. *Walking Methodologies in a More-than-Human World : WalkingLab*. Routledge. https://doi.org/10.4324/9781315231914.

Tsing, Anna. 2015. *The Mushroom at the End of the World: On the Possibility of Life in Capitalist Ruins*. Princeton University Press.

Wood, Dennis. 2010. *Rethinking the Power of Maps*. Guilford Press.
Yusoff, Kathryn. 2013. "Geologic Life: Prehistory, Climate, Futures in the Anthropocene." *Environment and Planning D: Society and Space* 31 (5): 779–95. https://doi.org/10.1068/d11512.
Yusoff, Kathryn. 2015. "Geologic Subjects: Nonhuman Origins, Geomorphic Aesthetics and the Art of Becoming Inhuman." *Cultural Geographies* 22 (3): 383–407. https://doi.org/10.1177/1474474014545301.

7

“My Place in Nature”

Significant Places in Nature and a Good Life

Jannicke Høyem

Human beings are always in relationship with the nature around them, but many factors influence how these relationships manifest and what meaning we assign to them. Memories and experiences of being outdoors are complex and bodily—tied to specific places, activities, and social relations. How do “the landscapes of childhood” and significant places in nature influence who a person is today?

This text takes as its point of departure an outdoor workshop I facilitated in Rosendal, Norway, in May 2022 (see also Peterson, this volume, Höckert et al., this volume). About twenty-five adults, all involved in the One by Walking research network, participated. Walking and talking in small groups, we explored memories of places in nature that were meaningful to us. The aim of this chapter is to convey how and why place is crucial in a time of climate crisis, while also offering an example of didactics for working outdoors with place.

Friluftsliv, Phenomenology, and Eco-Philosophy

To prepare for the workshop, before we met, I asked all participants to choose a place in nature that was meaningful to them, either one they remembered from childhood or one they had visited (perhaps regularly and close to home) as an adult.

In this work, I position myself within the field of *friluftsliv*. A Scandinavian term, *friluftsliv* means “outdoor recreation,” with an emphasis on free play, exploratory activities, place-based knowledge, local nature and cultures, and sustainability

(Gurholt and Sanderud 2016). *Friluftsliv* is understood as a diverse cultural and social phenomenon that describes human actions in and in relation to nature (Gurholt 1999; Tordsson 2003). In being outdoors, we interact and communicate with nature. Through this interaction, we discover not only nature but also each other and ourselves. Hence, children's outdoor play is a democratic place for learning about the world and about being human (White 2013). Our understanding of and affection for nature are often shaped by childhood encounters with the natural world—such as during outdoor play and other activities (Bischoff 2018; Fasting and Høyem 2022; Hendricks 2001). Encouraging engagement with nearby natural environments can foster a lifelong connection to the outdoors (Wattchow and Brown 2011). Høyem (2020) contends that *friluftsliv* can lay the groundwork for nurturing a conscious relationship with nature. Other research suggests that individuals with a strong connection to nature are more likely to act in environmentally friendly ways (Chawla 2020).

This work is grounded in the cultural phenomenological understanding of embodiment as something that not only belongs to the individual but also involves the interaction between the person, culture, and surroundings. The focus is on the interplay between culture, landscape, nature, and relationships, laying the foundation for the embodiment of experiences and memories. I understand the concept of memory as the revitalization of past moments, situated in time and place (Donohoe 2014), and preserved as embodied experiences (Casey 1997).

My Place in Nature

Rosendal in spring: The steep mountains, capped with snow, and the spring-green birches on the hillside above the fjord create a beautiful frame for an outdoor experience. The day before the workshop, I went for a run in the area to find a place well-suited for walking and talking in small groups, starting from the Barony.

The subjective, lived body is perpetually in dialogue with the world (Merleau-Ponty 1962), fostering connections between people and their environments. Experiences in nature can enhance individuals' relationships with the natural world throughout their lives. This consciousness of our own place is deeply intertwined with our emotional development and sense of self. Merleau-Ponty argues that the experience of existing in the world is something that people first and foremost have and do as a body. Through the body, the surroundings are sensed, understood, and translated into experience and consciousness (Merleau-Ponty 1962). An individual's experiences are framed by social and cultural contexts. An encounter with nature is not merely an isolated experience of being in a beautiful place but also an engagement with the location's history and uniqueness. In other words, the place and its inhabitants mutually influence each other (Beames et al. 2023, 60).

We have probably all experienced places that feel good for us—whether outdoors or elsewhere—and we may enjoy returning to some of them. It is hard to recall memories without also recalling where they took place. Conversely, it is difficult to think of a place without linking it to an experience there.

My first encounter with place theory was through the Norwegian architect Christian Norberg-Schulz (1992), and I was immediately fascinated. In my work with *friluftsliv*, I have often reflected on the significance of place. How can nature function as a place for learning, and which places in nature do we choose for outdoor education, therapy, rehabilitation, or recreation? And what is the interaction between a person and the outdoors like?

The concept of place is widely debated across multiple disciplines and understood differently by various researchers (Beames et al. 2023; Casey 1997; Cresswell 2004; Tuan 1979; Wattchow and Brown 2011; Wiestad 2014). My starting point here is the inspiration and insight I found in Norberg-Schultz. In his phenomenology of place, he emphasizes that life and the place where life is lived are inextricably connected, and that thoughts and feelings vary depending on one's location. Different places elicit different emotional and bodily experiences. Norberg-Schulz (1992) stresses that perception is a holistic bodily function in which all the senses participate and argues that phenomenology is a form of understanding that can deepen our awareness of human presence.

Norberg-Schulz (1993, 31) explains that maps are an important tool for understanding place, as they describe the natural givens that "shall provide room for human life," in line with the concepts of "location" and "locale" (Cresswell 2004)—the material setting of the place. However, a map only reflects the spatial organization of a place and presents a particular point of view, often compiled by those in power. A map cannot convey the character of a place. Lauvland (2007) argues that place must be decided both in terms of room and character. To describe the character of a place, or what Cresswell and Norberg-Schultz refer to as "sense of place," we need to supplement maps with other forms of description. The words we choose should articulate meaning, harmony, atmosphere, and mood. In this context, Norberg-Schultz also emphasizes the importance of being present, which he suggests is a qualitative experience: "to be connected to a place where the heart can bloom and the brain can play" (Norberg-Schulz 1996, 14). He describes this as a fundamental aspect of being human.

At the same time, the character of a place lies in the overall experience of it, so that both location and character are associated with what is collectively called "place." The landscape, as perceived by local people or visitors, evolves over time as a result of being acted upon by natural forces and human beings (Council of Europe 2000). A person develops their identity through knowing a place and taking care of it, by "being present" in accordance with the nature of the place and thus attaining a sense of place. Norberg-Schulz (1996, 18) refers to this as a "qualitative concept of place."

Morning light in the Barony Garden. The sky is blue, and the air is not yet warm enough to evaporate the dew on the grass after the cold of the night. The participants gather in a circle, shoulder to shoulder, to foster a sense of community. Each person gets to share the name and location of their chosen place in nature before I offer an introduction, in line with the text above.

Within the field of *friluftsliv*, the concept of place phenomenology—inspired by Merleau-Ponty, Casey, and Norberg-Schulz—has been widely used in studies of humans' relationship to nature and the connections between place and subjectivity. What is central in *friluftsliv* is active human interaction with nature. Through our actions in the world, we experience both its properties and ourselves. Different tasks organize the world in different ways and appeal to various aspects of our being. To describe our place within this dynamic, both Merleau-Ponty (1962) and Norberg-Schulz (1992) employ the phenomenological idea that humans "inhabit" the world. Place is understood as a phenomenon and a dimension rather than a mere geometrical space. A similar perspective on humans' relationship to nature can be found in the ecosophy of Næss (1974), who promotes a nonanthropocentric view that emphasizes the mutual connection and interdependence between humans and nature. Thus, both phenomenology and ecosophy contribute to a holistic understanding of human-nature relationships.

Casey (2001, 684) ties place and self together, stating that "there is no place without self and no self without place." Norberg-Schulz (1992) illuminates the phenomenological concept of space by using the term "genius loci": the spirit of place. Tuan (1979) distinguishes between space and place, with space allowing for movement, while place signifies a pause in movement, enriched by familiarity and appreciation. However, Abram (1997) underscores the dynamic nature of place, even in its stillness, attributing it to the potential for experience it contains. Consequently, nature can be viewed as space when it serves merely as a backdrop, and as place when a connection is made with it.

The Landscapes of Childhood

Following the introduction, the participants were divided into groups of three. Setting out past the entrance of the Barony, they continued along the small, unpaved road beside the river on the northern side for about ten minutes, until they reached the next meeting point. As they walked and talked, each group member was invited to present their chosen place in nature through its space, character, and social and bodily memories. The group then collectively reflected on how these places were integral to their own identities and were encouraged to offer richer, more detailed descriptions.

The concept of landscape encompasses the interaction between sociocultural and physical-material environments, how people experience them, and how meaning is constructed through encounters with the landscape, shaped by cultural and individual meaning systems as well as the landscape itself (Ingold 2000; Bischoff 2012; Rosa 2019). Building on this, and drawing on Fasting et al. (2022), I refer to the participants' chosen places in nature as "landscapes of childhood." In this context, the concept of landscape incorporates the significance of nature as a place in relation to the self—through memories, meaning-making, and embodiment—and as part of one's identity.

Sandberg and Vuorinen (2010) highlight that our recall of the past is mediated by the ideas, thoughts, and experiences we accumulate later in life, suggesting that adults tend to idealize their childhood. Berntsen (2014) argues that we do not necessarily remember past events accurately; rather, our memories are largely retellings colored by our present circumstances and the interpretations we impose on them. Research indicates that the significance we attribute to encounters with nature is predominantly formed during childhood (Bischoff 2012, 2018; Tuan 1979; Sandberg 2003), a process further reinforced through schooling and education.

What Makes a Place in Nature Meaningful to Us?

At the final meeting point, each group selected one of their places to present. Several groups noted how they had identified similarities in the different places chosen by the members, despite their diverse backgrounds, nationalities, education, gender, and age. For example, a lighthouse, a beach, and a creek provided common ground for exploring one's relationship with water. Through these conversations, it became clear that places in nature were often linked to identity, a sense of belonging, and embodied and social experiences—impressions that resonate with place theory.

Identity

Wiestad (2014, 25) posits that the imprints left by place experiences, including habits, skills, and memories, "become part of our self-experience" over time. Places contribute to shaping our self-perception by anchoring us (Fasting and Høyem 2022). We are anchored in the landscape through memories tied to specific places and actions (Ingold 2000; Casey 1997; Norberg-Schulz 1996). Place identity, shaped by being in the landscape, can foster a sense of belonging, anchoring, and resonance in a changing world and seems to be part of what is experienced as meaningful.

In today's rapidly changing environment, cultivating a sense of stability and personal identity amid social transformations is crucial. This feeling often arises in places where we experience a sense of connection and familiarity (Corbett 2006). As individuals engage repeatedly with a location's historical, scientific, and aesthetic elements, their connection to it can deepen and solidify (Wattchow and Brown 2011).

Belonging and Resonance

The interplay between a location and human history is pivotal to our perception of the place and our relationship with it (Wattchow and Brown 2011, 62). This remains true even when we have not personally experienced the history of the location but have only heard it retold. One might argue that there is a universal human quality to the interaction between nature and humans that binds individuals across time. Places help shape our self-perception by anchoring us through memories tied to specific locations and actions. This transformation from space to place facilitates a deeper bond with nature, moving beyond its depiction as merely a backdrop for human activities. In a place-based approach, nature transitions from being perceived as a space to being recognized as a place, fostering a sense of anchoring and encouraging individuals to see themselves as integral components of the natural environment, rather than as separate entities.

Places in nature constitute a physical, social, and cultural context for experiences and encounters. During the workshop in Rosendal, the relationship between the individual and the place became evident as memories were recalled, manifesting in the resonances that arose between the participants and the places they had chosen. The notion of "resonance" builds on Rosa (2019), who examines individuals in a world characterized by increasing velocity (physical, spiritual, social, and existential), developing resonance in contrast to acceleration. Resonance should be understood through a phenomenological lens of human attention and intentionality as a reciprocal relationship: the subject is touched by the world and, in turn, touches the world through action (Rosa 2019). This reciprocity nurtures wonder and teaches the individual to approach the world with openness (Fasting and Høyem 2022). The participants' stories about their places in nature exhibited resonance in Rosa's sense. Resonance refers to how, for instance, recalling a place can evoke a particular mood and how a single impression in that place can trigger a range of emotions and activate memories. The concept illuminates the qualities of the relationship between a person and a place. In Rosendal, many participants described memories of places in nature with which they had connected during childhood or adolescence. Others chose places that resonated with them later in life, and my impression is that these places resembled the landscapes of their childhood.

Embodied Experiences

The stories told in Rosendal recalled movement—walking, running, jumping—and sensory memories—smells, sounds, and tactile impressions. One participant described the smell of sunshine on heather in the spring in his place. Casey (1997) illustrates how bodily memories create a sense of belonging to places. Through the body's senses, relationships with our surroundings are formed in a continuous stream of impressions. As sensing bodies in motion, we inhabit the world in a phenomenological sense, and our surroundings become part of us (Casey 1997; Merleau-Ponty 1962). The body carries traces of the places we have known, and memories of these places are stored as embodied experiences (Casey 1997; 2001). This aligns with the research of Österlund-Pötzsch (2023), who, instead of tying the concept of "home" to a geographical location, encourages us to see it as a place created through our practices, rhythms, attitudes, and thoughts. She emphasizes how childhood encounters with the local natural environment are often formative in evoking a sense of home.

Social Experiences

Nature is a suitable arena for social interaction (Pálsdóttir et al. 2018). Social relationships are crucial for experiencing quality of life and wholeness, and for having confidence in one's ability to handle life's ups and downs (Roberts et al. 2020). We often experience being in nature with others. In Rosendal, several workshop participants emphasized the social dimension of their chosen places. This was exemplified by the story of a little girl who used to walk in the local hills with her grandfather and, now as an adult, recollected their conversations and relationship while hiking the same paths.

Fasting et al. (2023) found that being in nature creates different conditions for social interaction than those experienced indoors. The context of nature plays a significant role in the experience of social belonging. Nature treats everyone equally. Their research suggests that nature offers a sense of freedom from the societal pressures of comparison, providing a platform for community and social interaction that limits these mechanisms.

My Place in Nature in the Face of Today's Challenges

In today's world, access to blue and green spaces varies (EEA 2023), and children often have limited opportunities for free outdoor play and consistent interaction with nature (Beery et al. 2023; Haidt 2024; Louv 2005; Lumber et al. 2017). Experiences of being outdoors are believed to have significant impacts, yet, for many, a childhood of wandering and exploring has been replaced by adult supervision (Chawla 2015; Mygind et al. 2018. This escalating disconnection from—and unfamiliarity with—places leads to placelessness or rootlessness

(Baker 2005). This is driven by factors such as increased mobility, weakened community ties, economic globalization, environmental degradation, and alienating architecture (Kellert 2005, 60).

To counteract this trend, a shift toward appreciating the specifics of a place—rather than abstract notions of nature—is needed (Wattchow and Brown 2011). The concept of "place responsiveness" seeks to bridge this gap (Mikaels 2018). This approach considers the historical, present, and potential future uses of places (Beames et al. 2023). A place serves as a source of rootedness, offering a sense of security, belonging, and stability. By giving participants an opportunity to recall their relationships to places in nature that were important to them, the workshop in Rosendal facilitated place responsiveness.

Fasting et al. (2022) demonstrate that young adults' bodily exploration during outdoor activities is fundamental in shaping what they call "the landscapes of childhood." Memories of these places impact who they are today and form the foundation for how they relate to nature and their fellow human beings. Place identity, often shaped through interactions with childhood landscapes, can offer a sense of belonging, anchoring, and resonance in a world that is constantly evolving. This resonates with the conversations the workshop participates had while walking outside the Barony in Rosendal.

Individuals who develop a strong attachment to a specific place are more likely to exhibit nature-conscious and environmentally friendly behavior (Chawla 2020; Corbett 2006). However, the growing disconnection of contemporary society from the environments it inhabits can foster a rising sense of placelessness (Gray 2005). Such feelings of estrangement can diminish people's commitment to preserving the cultural and ecological features of their surroundings (Kellert 2005).

Wattchow and Brown (2011) delve into the true meaning of the term "place," noting that modern humans have, in many ways, detached themselves from their natural position in the environment, often surrounding themselves primarily with other humans or human-made objects. They argue that when we lose our connection and reciprocal dialogue with nature, something essential is lost. Wise et al. (2023, 238) claim that our current ecological crisis "is nothing more than a crisis of relationship." Wattchow and Brown (2011) emphasize the value of raising awareness about the interconnectedness between the place where we find ourselves and the rest of the world, while Lynch and Mannion (2021) describe cultivating a sensitive and sensible interaction between people and their environments as a key principle for developing sustainable perspectives. Recalling "my place in nature" gave the participants in Rosendal an opportunity to heighten this awareness, and the workshop highlighted how this can contribute to a good life for both people and the planet. We all live on the same planet. Borders are constructed by humans—and nature doesn't care about them. Modern society values novel experiences and mobility, but a solid understanding of the local is crucial for making sense of new places and contexts (Wattchow and Brown 2011). A place can serve as a link between the local and the global, connecting us not only to a specific location but also to ourselves, to others, and to broader ideas of nature.

References

Abram, David. 1997. *The Spell of the Sensuous: Perception and Language in a More-than-Human World*. Vintage Books.

Baker, Molly. 2005. "Landfullness in Adventure-Based Programming: Promoting Reconnection to the Land." *The Journal of Experiential Education* 27 (3): 267–76. https://doi.org/10.1177/105382590502700306.

Beames, Simon, Peter Higgins, Robbie Nicol, and Heidi Smith. 2023. *Outdoor Learning Across the Curriculum: Theory and Guidelines for Practice* (second edition). Routledge. https://doi.org/10.4324/9781003010890.

Beery, Thomas, Anton Stahl Olafsson, Sandra Gentin, et al. 2023. "Disconnection from Nature: Expanding Our Understanding of Human-Nature Relations." *People and Nature* 5 (2): 470–88. https://doi.org/10.1002/pan3.10451.

Berntsen, Dorthe. 2014. *Erindring*. University Press.

Bischoff, Annette. 2012. "Mellom meg og andre er der stier. En avhandling om stier, mennesker og naturopplevelser." PhD diss., Norwegian University of Life Sciences.

Bischoff, Annette. 2018. *Betydningen av barndommens landskap i voksen alder*. https://norskfriluftsliv.no/wp-content/uploads/2019/06/Forskning-i-friluft-2018-Konferanserapport.pdf.

Casey, Edward. 1997. *The Fate of Place. A Philosophical History*. University of California Press.

Casey, Edward. 2001. "Between Geography and Philosophy: What Does It Mean to Be in the Place-World?" *Annals of the Association of American Geographers* 91 (4): 683–93.

Chawla, Louise. 2015. "Benefits of Nature Contact for Children." *Journal of Planning Literature* 30 (4): 433–52.

Chawla, Louise. 2020. "Childhood Nature Connection and Constructive Hope: A Review of Research on Connecting with Nature and Coping with Environmental Loss." *People and Nature* 2 (3): 619–42. https://doi.org/10.1002/pan3.10128.

Corbett, Julia B. 2006. "The Formation of Environmental Beliefs." In *Communicating Nature: How We Create and Understand Environmental Messages*. Island Press. http://choicereviews.org/review/10.5860/CHOICE.45-2682.

Council of Europe. 2000. European Landscape Convention, ETS 176. https://www.coe.int/en/web/landscape/the-european-landscape-convention.

Cresswell, Tim. 2004. *Place: A Short Introduction*. Blackwell.

Donohoe, Janet. 2014. *Remembering Places: A Phenomenological Study of the Relationship between Memory and Place*. Lexington Books.

European Environment Agency (EEA). 2023. *Who Benefits from Nature in Cities? Social Inequalities in Access to Urban Green and Blue Space across Europe*. EEA Report No 21/2019. European Environment Agency.

Fasting, Merete Lund, and Jannicke Høyem. 2022. "Freedom, Joy and Wonder as Existential Categories of Childhood—Reflections on Experiences and Memories of Outdoor Play." *Journal of Adventure Education and Outdoor Learning* 24 (2): 145–58. https://doi.org/10.1080/14729679.2022.2066008.

Fasting, Merete Lund, Jannicke Høyem, and Annette Bischoff. 2022. "Barndommens landskap. En studie av unge voksnes minner i møte med barndommens nærnatur." *Journal for Research in Arts and Sports Education* 6 (1). https://doi.org/10.23865/jased.v6.3017.

Fasting, Merete Lund, Jannicke Høyem, Cathrine Neresen, and Bjørn Tore Johansen. 2023. "Friluftsliv og psykisk helsearbeid: Hvordan sosialt friluftsliv, bevegelsesglede og naturnærvær kan bidra til livsmestring." *Tidsskrift for psykisk helsearbeid* 20 (3): 174–86. https://doi.org/10.18261/tph.20.3.2.

Gray, Tonia. 2005. "Exploring Our Connections and Relationships with Place and/or Nature." In *Outdoor and Experiential Learning: Views from the Top*, edited by Tracey J. Dickson, Tonia Gray, and Bruce Hayllar. Otago University Print.

Gurholt, Kirsti Pedersen. 1999. "'Det har bare vært naturlig!' Friluftsliv, kjønn og kulturelle brytninger." PhD diss., Norwegian School of Sports Sciences.

Gurholt, Kirsti Pederson, and Jostein Rønning Sanderud. 2016. "Curious Play: Children's Exploration of Nature." *Journal of Adventure Education and Outdoor Learning* 16 (4): 318–29. https://doi.org/10.1080/14729679.2016.1162183.

Haidt, Jonathan. 2024. *The Anxious Generation. How the Great Rewiring of Childhood Is Causing an Epidemic of Mental Illness*. Penguin Books.

Hendricks, Barbara. 2001. *Designing for Play*. Ashgate.

Høyem, Jannicke. 2020. "Outdoor Recreation and Environmentally Responsible Behavior." *Journal of Outdoor Recreation and Tourism* 31: 100317. https://www.sciencedirect.com/science/article/pii/S2213078020300414.

Ingold, Tim. 2000. *The Perception of the Environment: Essays on Livelihood, Dwelling and Skill*. Routledge.

Kellert, Stephen R. 2005. *Building for Life: Designing and Understanding the Human-Nature Connection*. Island Press.

Lauvland, Gro. 2007. *Verk og vilkår: Christian Norberg-Schulz' stedsteori i et arkitekturfilosofisk perspektiv.* PhD diss., Oslo School of Architecture and Design.

Louv, Richard. 2005. *Last Child in the Woods: Saving Our Children from Nature-Deficit Disorder.* Algonquin Books.

Lumber, Ryan, Miles Richardson, and David Sheffield. 2017. "Beyond Knowing Nature: Contact, Emotion, Compassion, Meaning, and Beauty Are Pathways to Nature Connection." *PLos ONE* 12 (5): e0177186. https://doi.org/10.1371/journal.pone.0177186.

Lynch, Jonathan, and Greg Mannion. 2021. "Place-Responsive Pedagogies in the Anthropocene: Attuning with the More-than-Human." *Environmental Education Research* 27 (6): 864–78. https://doi.org/10.1080/13504622.2020.1867710.

Merleau-Ponty, Maurice. 1962. *Phenomenology of Perception.* Routledge & Kegan Paul.

Mikaels, Jonas. 2018. "Becoming a Place-Responsive Practitioner: Exploration of an Alternative Conception of Friluftsliv in the Swedish Physical." *Journal of Outdoor Recreation, Education, and Leadership* 10 (1): 3–19. https://doi.org/10.18666/JOREL-2018-V10-I1-8146.

Mygind, Erik, Mads Bølling, and Karen S. Barfod. 2018. "Primary Teachers' Experiences with Weekly Education Outside the Classroom during a Year." *International Journal of Primary, Elementary and Early Years Education* 47 (5): 599–611. https://doi.org/10.1080/03004279.2018.1513544.

Næss, Arne. 1974. *Økologi samfunn og livsstil.* Universitetsforlaget.

Norberg-Schulz, Christian. 1992. *Mellom jord og himmel: En bok om steder og hus.* Pax.

Norberg-Schulz, Christian. 1993. *Nattlandene: om byggekunst i Norden.* Gyldendal.

Norberg-Schulz, Christian. 1996. *Stedskunst.* Gyldendal.

Österlund-Pötzsch, Susanne. 2023. "Hem, Hemlös, Hemlig: Reflektioner kring vandrande och spatial tillhörighet—ett metodologiskt experiment." *Budkavlen* 92: 40–49. https://doi.org/10.37447/bk.129758.

Pálsdóttir, Anna Maria, Joe Sempik, Willian Bird, and Matilda van den Bosch. 2018. "Using Nature as a Treatment Option." In *Oxford Textbook of Nature and Public Health: The Role of Nature in Improving the Health of a Population*, edited by Matilda van den Bosch and William Bird. Oxford University Press.

Roberts, Anna, Joe Hinds, and Paul M. Camic. 2020. "Nature Activities and Wellbeing in Children and Young People: A Systematic Literature Review." *Journal of Adventure Education and Outdoor Learning* 20 (4): 298–318. https://doi.org/10.1080/14729679.2019. 1660195.

Rosa, Hartmut. 2019. *Resonance: A Sociology of Our Relationship to the World.* Polity Press.

Sandberg, Anette. 2003. "Play Memories and Place Identity." *Early Child Development and Care* 173 (2–3): 207–21.

Sandberg, Anette, and Tuula Vuorinen. 2010. "Reflecting the Child: Play Memories and Images of the Child." In *Engaging Play*, edited by Liz Brooker and Susan Edwards. Open University Press.

Tordsson, Bjørn. 2003. *Å svare på naturens åpne tiltale: En undersøkelse av meningsdimensjoner i norsk friluftsliv på 1900-tallet og en drøftelse av friluftsliv som sosiokulturelt fenomen*. PhD diss., Norwegian School of Sports Sciences.

Tuan, Yi-Fu. 1979. "Space and Place: Humanistic Perspective." In *Philosophy in Geography*, edited by Stephen Gale and Gunnar Olsson. Springer Netherlands. https://doi.org/10.1007/978-94-009-9394-5_19.

Wattchow, Brian, and Michael Brown. 2011. "The Emergence of Place in Outdoor Education." In *A Pedagogy of Place: Outdoor Education for a Changing World*, edited by Brian Wattchow and Michael Brown. Monash University.

White, Jan. 2013. *Playing and Learning Outdoors*. Routledge.

Wiestad, Else. 2014. *Stedene som former deg. Om rom og subjektivitet*. Fagbokforlaget.

Wise, Maggie, Bruce Martin, Andrew Szolosi, and Tamarine Foreman. 2023. "Habits of Connection: From Sustainability and Saviorship to Reciprocity and Relationship." *Journal of Experiential Education* 46 (2): 238–55. https://doi.org/10.1177/10538259221134876.

8

Walking Is Also Space

Walking Beyond Distance

Ami Skånberg and Anna Viola Hallberg

Introduction

"Walking Is Also Space" was an extension of Skånberg's ongoing walking practice within choreography, investigating space with an emphasis on experiential and inclusive aspects. Skånberg and Hallberg worked together 2023-2025 and in this workshop, performed in Finland 2023, they integrated methodologies from lens- and embodied practices. However, when considering the universal walker/wayfarer, the focus is often on the distances they cover and overcome. Distance equals time, with spaces—stereotypically new territories or very challenging trails—seen as obstacles for walkers to conquer. The aim of this workshop was to raise awareness about disregarded or unfamiliar spaces, contributing to solutions for overwalked trails. This meant creating situations where a walker's goal was not to cover distance but to experience proximity—where the internal world could encounter and share both real and imagined spaces. The workshop has continued to function as a methodology for attentive walking projects performed and cocurated in Sweden and internationally.[1]

For the creation of empathetic trails, we argue for the need to pay attention to both participants and space/nature in order to reduce the stress on trails. We framed the outdoor/indoor workshop through a methodology of "share sessions," where the group—not the leader—becomes the knowledge holder. This is a method for informal sense-making and creating experiences of togetherness. We divided the participants into pairs, each with a different assignment: one was to be the Walker, the other the Observer. There were six Walkers, six Observers, and we also assigned a specific task to one of the participants: the Mapper who would make a bird's-eye view drawing (see Figure 8.1). Within the artistic research process, documentation was carried out through both participants' self-recorded footage, and our own video recordings of the workshops and sharing sessions.[2]

The practice-led research study drew on ethnographic methodologies commonly used in artistic research. What connects ethnography and artistic research in this context is societal engagement—a curiosity about the social interventions and perceptions that occur between us and the environment. The aim was to evoke the imaginary and raise awareness of how we position ourselves together in the world though the embodied memories engendered by movement, along with the collection of detailed observations and interviews. The reflections in this text are based on examples from our experiences and those of the participants.

Doreen Massey (2005) claims that spaces are never static; they transform as we move through and within them. Spaces are never empty because we always bring both our immaterial and material knowledge with us. In this sense, walking is also space. The practice-based content of our workshop was inspired by choreographer Molissa Fenley's ideas about space, which Skånberg studied at the American Dance Festival in 1997.[3] Fenley's experience with Nigerian Yoruba dances included practices that create empathic involvement, where one witnesses and becomes absorbed in something to the point that time seems to disappear. These dances involved both dancers and observers, which inspired Fenley to integrate this structure into her own choreographic and teaching practices (Williams 2013).

In the workshop, we therefore assigned participants the roles of Walkers and Observers. We saw the potential to translate an empty "canvas" into a representation of nontheatrical spaces—in this case, an "empty," abandoned field in Korpoström. In this field, we applied Fenley's choreographic practice to move away from dominant diagonals, solo positions, and dramatic perspectives. Our question was: Could the workshop support the creation of empathetic trails by helping participants find and appreciate the gaps, the movements within them, thereby reducing stress on both the trails and the walkers?

Walking Score for "Walking Is Also Space"

"Walking Is Also Space" was organized in three phases, each with its own instructions:

1. Introduction
2. Walk/Choreographed Stories
3. Share Session

The workshop was documented through cameras, one drawing, and notes taken by both the group and us, which we revisited during the writing of this text. Although the participants were assigned different roles and tasks as Walkers and Observers, they shared the same space. However, because each had their own register and reference system, they experienced the surroundings and atmospheres differently. In addition, one person was assigned the role of Mapper, tasked with taking a bird's-eye view of the workshop and drawing a map of the various walks in the space. Neither the participants nor we had any prior knowledge of the setting in Korpoström.

Instructions for Observers

The Observers gathered indoors to hear the instructions and share their experiences:

- Pay attention to both passive and active information gathering, as this will later be decoded and compared.
- Familiarize yourself with the recording setup—the lay of land, weather conditions, etc.—so you can focus on the Walker.
- Hold your camera near the lower part of your torso and direct it at the Walker's torso, not at the landscape.
- If it is windy, and you don't have a windshield to put over the microphone, use a mitten, sock, or something similar that is loosely threaded to cut the wind.
- Practice using the on/off buttons and zoom on your camera, so you can operate it without looking at the screen and/or control buttons.
- Prepare to keep your eyes on your Walker (and on the uneven field so you don't trip).
- Please do not talk during the process.
- Please stay clear of other Walkers and Observers (both on camera and on the field).

Instructions for Walkers

While the Observers received their instructions, the Walkers were given others. The group of six Walkers, along with the bird's-eye view Mapper, gathered outdoors and proceeded to a nearby field together. Following Molissa Fenley's ideas about space, the Walkers were asked to select a point anywhere on the field and name it "Point One." This process was repeated several times, progressing slowly, until all six Walkers had chosen four different points, creating five individual paths between the points: starting at Point One, advancing to Two, Three, and Four, and ultimately returning to Point One. The paths between the points were walked and rehearsed four times. The task proceeded slowly to allow more sensory information about the abstract geographical points to be gathered, with the support of the following cues:

- Proceed slowly, constructing the four points/spots, which you will return to.
- Point One becomes a place. Do you remember any warm places from your life? Stay there and experience the warmth.
- Point Two becomes a place. Now walk toward Point Two, which is a windy place. Do you remember any windy places from your life? How do you move from a warm place to a windy place? Stay there and experience the wind.
- Point Three becomes a place. Now walk toward Point Three, which is a silent place. Do you remember any silent places from your life? How do you move from a very windy place to a silent place? Stay there a moment and experience the silence.
- Point Four becomes a place. Continue to Point Four, which is a very muddy place. Do you remember any muddy places from your life? How do you move from a silent place to a muddy place? Stay there and experience the mud.
- Now return to Point One, the warm place. How do you move from a muddy place to a warm place? Stay there and experience the warmth.
- Depart from your personal point at the same time, sensing the space and the other participants.
- Arrive at your personal point in your own time.
- Wait until all Walkers have reached their personal points before proceeding.
- Repeat your created paths three times in the presence of the Observers.

By following these instructions, participants created an organic, choreographic space, with some Walkers moving while others waited in stillness.

Instructions for the Mapper: The Bird's-Eye View

One participant was tasked with documenting what happened on the field by creating an overall "score" in the form of a drawn map. As a meta-observer, the

Mapper noted rhythms, signs, and the parameters of the field, according to the instructions indicated here:

- Pay attention to and draw both the narrative of the choreography.
- Pay attention to and draw notations for the "vulnerable"/trails created on the field.
- Take notes by drawing the points, paths, and cues of all participants.
- Include spontaneous interpretations of what might take place between Walkers and Observers as they proceed.

By following these instructions, the Mapper created a notation of the choreography as it evolved in the field (see Figure 8.1).

The Walk: Observers, Walkers, and the Mapper Come Together

The Observers began their walk toward the field, trying out camera settings while being in motion. Upon arriving, the Observers were paired with their Walkers. Next, the Observers lined up at the edge of the field beginning their observation from afar with the wind in their backs. To work with both distance and proximity, the Observers' first task was to look at "their" Walker, that is, follow the Walker from afar. All Observers lined up along the far edge of the grass field to track their respective Walker, taking written notes. The Observers' cue was to be with their Walkers as silent witnesses. For the second repetition, the Observers documented the Walkers' paths with their cameras still remaining in the lined-up position. For the third, they moved in, to film closer to "their" Walker. This progression allowed the Observers to first watch the Walkers' walking choreography from a distance, then document their paths with cameras from afar, and finally actively follow the Walkers up close. These various perspectives—distance and proximity—allowed for both passive and active information gathering.

The Walkers choreographed themselves individually—responding to both the field and their own memories, merging their inner and outer landscapes. By this, they established their individual movements and rhythms. The Walkers were instructed to wait for each other at each of the four key points, due to the durational variations of the individual paths. Where paths overlapped, Walkers were respectful and attentive to avoid collisions or interference, creating a distinct rhythm. This attentiveness was heightened when the Observers entered the field.

The Mapper's observation and documentation resulted in a drawing of the movements performed by both the Walkers and the Observers. The Mapper's meta-observation added an invaluable contribution, sensing and noting all rhythmic patterns at the same time. The space itself became a field of relational walking research, interpreted sensitively by the Mapper from a bird's-eye view. Following the process closely, almost obsessively, the Mapper noticed how the walk-paths evolved, without a fixed script or map.

Share Session: Storying Space

Indoors, the pairs began by walking around the gallery, exchanging experiences from the field. We then formed a circle, with each pair sitting together, various narratives and insights emerging. The material collected by both the Observers and the Mapper was decoded and compared with the stories shared by the Walkers. The Mappers presented their drawn notations of movements, stick figures, and landscapes, as well as some of the instructions they had devised. The Observers and Walkers highlighted how the Mapper's presence as a witness/audience gave the experience a sense of significance. The Walkers noted that being observed was crucial—it made the situation feel special, as there was a receiver or witness for their choreographed walk-paths. The Observers, on the other hand, realized the importance of their own role, describing how the workshop created an unexpectedly strong sense of empathy and care for their assigned Walker.

The different modes of observation, influenced by the physical proximity of engagement, were also explored. For instance, Observers felt a deep concern for the Walkers' well-being when moving closely alongside them; this proximity evoked a strong sense of responsibility, making the Observers feel they would do anything to support their Walker. In contrast, when observing from a distance and taking notes, their attention shifted to entirely different aspects of the experience.

The Mapper was intrigued by the dramaturgical development of the walk-paths: the meticulous evolution from the initial selection of the individual geographical points, devoid of content, to the subsequent embodiment of those points through the cues. The Walkers agreed, reflecting on how easily they were

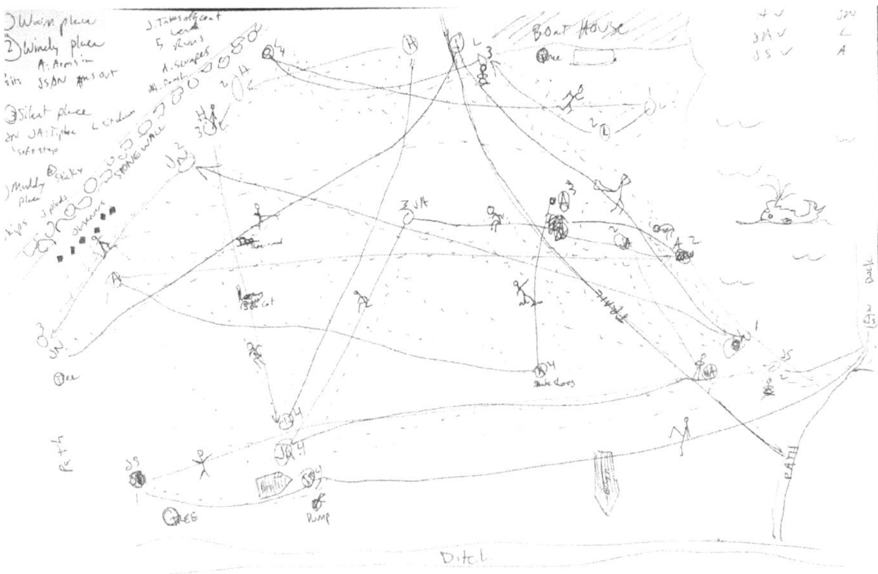

Figure 8.1 The Mapper's drawing of the uneven grass field in Korpoström, which was next to the Baltic Sea. To highlight this, the Mapper drew a whale. Whales are very rare in these waters. Source: Workshop-created material.

able to repeat their choreographed walk-paths on the field. They attributed this to the preassigned geographical points and, even more so, to the embodiment of personal memories triggered by the cues. It became clear that, even when the Walkers and the Observers were in the same place at the same time, their experiences diverged due to their individual body memories, registers, and reference systems. There was an emphasis on both actual events and interpreted experiences. The cues inspired the Walkers to make creative choices: flying like a crane, arriving in the reefs, or struggling against the wind on a hill. Some Walkers noted that, in order to support their personal narrative and make themselves comprehensible to the Observer, they felt compelled to emphasize their story through exaggerated movements. This raised the question of how much effort we put into being understood by others in society. Could we more fully appreciate the extra work of care performed by our coresearchers, both at work and in everyday life?

One Walker explained that their decision to imagine animals through movement made the task more manageable. Another Walker found that the choreographic cue "windy place" suddenly transformed into a monsoon in another part of the world, sparking yet new stories. For some, the field's gusts of wind triggered climate awareness; for others, they evoked childhood memories or a sense of safety. The imagined wind from elsewhere "collaborated" with the real wind at the site. The discussion also touched on the pros and cons of the various note-taking approaches and how the focus of attention varied between them.

The workshop did not demand a singular interpretation of what was happening on the grass field. For example, one Observer interpreted all their Walker's movements as childlike, infused with childhood memories, even though the Walker had no intention of acting like a child. The workshop created a playfulness and intimacy in the dialogues, perhaps because it allowed for a new way to get to know and understand each other and our different perspectives on this shared experience.

Conclusion

Writing about the workshop helped us recognize and identify the tacit knowledge embedded in the participants' choreography. Embodied and spatial details, though implicit, were essential for creating a dedicated presence with others. The core idea behind "Walking Is Also Space" was to allow the choreographed walk-paths to inspire personal stories, with the actual site itself becoming the foundation for the creation of empathetic trails and acts of care. The workshop unpacked (re-, de-, un-)conditional walking through spatial agency, personal memory, and responsibility. We found that the self-explorative and unconstrained nature of our workshop created a deeper understanding of one another, both as persons and as researchers. While the shared interest in walking brought everyone together, the starting points varied widely, shaped by each participant's unique situation, research focus, and concerns (the global climate crisis, the destruction caused by wars and conflicts, the unreal and the real world, various forms of activism, and so on). The workshop encouraged participants to discover and value the gaps, the movements within the creation of empathetic trails. This complemented fixed, static trails, which often emphasize conquering distance. We found that conditional walking contributed to a more sustainable practice, thus reducing stress on both the trails and the walkers. We argue that this practice of care and sensitivity opens up contexts where we can create a sense of uniqueness, even when treading on disregarded and unfamiliar areas.

"Walking Is Also Space" engaged spatio-temporal narratives, interweaving and enhancing them with personal walking testimonials. Each participant brought their own chosen perspective to the space, which choreographed the layering of stories. This resulted in stories that extended beyond the verbal into a shared, embodied experience and understanding. The workshop also provided an alternative to boastful narratives about long and challenging walks, offering something different from a culture based on admiring achievement. Instead it revealed the possibility of finding meaning even in a space that might otherwise be categorized as dull. This opens up new perspectives on how to approach empathetic trails with care in the future. An unexpected outcome of the workshop was the transformation of group dynamics. The process accelerated participants' ability to get to know each other, allowing them to connect their own walking research to personal and political topics. People shared vulnerable matters from their lives that they might not have disclosed otherwise, creating a deeper sense of closeness and understanding within the group.

Notes

1 http://walkingaspractice.org/research.
2 The original plan was to create a multicamera documentation of the workshop and to highlight the notation differences between video and text.
3 https://americandancefestival.org/about/.

References

Massey, Doreen B. 2005. *For Space*. Sage Publications.

Williams, Cynthia. 2013. "Searching the Darkness: Molissa Fenley's *State of Darkness*." Sacre Celebration, 2013. https://sacre.info.yorku.ca/files/2013/10/Williams.pdf ?x31496.

9

Walking Back Waste

Rethinking Transects to Challenge Human Nature in Nature

Roger Norum and Rose Keller

It is a chilly September morning in western Iceland. Two dozen early risers are wandering through the tussocky landscape. A few of them eagerly make a beeline for a steaming geothermal spring pool at the edge of the plot of land. Shedding their robes, they immerse their weary bodies in the warm soup of organics and minerals. So begins a slow but steady flow of visitors to this oasis of calm, with varying volumes of activity throughout the day. At first glance, it may seem that most of these visitors, who traipse about western Iceland's landscapes and bathe in its springs, leave behind nothing but the shoe-shaped traces of their footsteps in the dirt, mud, and grass along the uneven paths. But as the number of visitors to remote areas continues to rise—and as infrastructure expands to accommodate them—humans will inevitably leave other traces on both the landscape and the social fabric of local communities. Among these marks, one particular remainder of human presence is currently terraforming this land and altering human-environment and place-based relations. This chapter explores the methodological approach of an intervention that highlights group practices and experiences related to the excrement of tourism.

The problem of open defecation and the accumulation of fecal matter in national parks, as well as in protected and natural recreation areas, is becoming increasingly prevalent worldwide. This is largely due to the rise in nature-based tourism, which currently accounts for approximately 7 percent of tourism globally, a figure considerably higher in the European Arctic (WTTC 2022; NHO 2025). As a result, those involved in developing and managing nature areas popular with tourists are seeking sustainable solutions to mitigate the accumulation of human waste, while also addressing the high costs of building and maintaining waste management facilities. We argue that the growth in nature tourism necessitates proactive measures, increased funding for the management of protected lands, and collaborative efforts between park authorities, visitors, and tourism business to manage the problem of human waste. Educating visitors on the environmental impact of improper waste disposal, promoting "Leave No Trace" principles, and providing access to tools for waste disposal can encourage responsible behavior (NCBI 2021).

In this chapter, we explore the educational insights generated when individuals are compelled, through in situ walking methods, to envision the en masse encounter with waste in spaces typically imagined (or marketed) as pristine. We examine how conservation policies often promote the idea of untouched, free nature while simultaneously regulating the corporeal mobilities allowed within it. Additionally, we investigate how the request for individuals to carry their own waste to mitigate structurally driven environmental damage from mass tourism may conflict with the allocation of resources for visitor management, prompting a critical assessment of what this means for practices in nature-based tourism.

To address these three issues, we begin by describing an intervention held during a One by Walking network "walkshop" in September 2023 in Sælingsdal, Iceland. The aim of the intervention was to illustrate how fundamental human practices can become exponentially problematic as the number of humans increases in time and space. Next we discuss the challenges and outcomes of the intervention, examining the paradoxes inherent in nature conservation by unpacking individual subjectivities and internalized controls, as revealed through reflections on participants' encounters with waste. Finally, we reflect on what happens to practices of walking in nature when visitors must truly leave *no* trace behind, when the "good tourist" becomes a mobile body that carries in and out everything they bring, disposing of excrement in another place and at another time. This intervention sought to introduce, at a smaller scale, an activity piloted during a three-year study in Lofoten, Norway, which explored whether tourists could adapt to carrying their own feces in specially designed waste bags (see Keller and Norum, this volume; Keller et al. 2025).

Encountering Waste at World's End

When studying environmental futures, social scientists employ a range of methodological approaches to structure inquiry, guide data collection and decision-making, and envision alternative pathways. These include tactics such as living labs (Lapointe et al. 2021), social learning (Andrade et al. 2023), PPGIS (Korpilo 2023), and citizen science more generally (Haklay et al. 2021), all of which seek to scientifically engage local actors in the environment. Another method is scenario modeling, a technique for both data collection and analysis that involves creating and analyzing potential futures based on assumptions about what is likely to happen.

A specific approach to scenario modeling, often used in sustainable development planning, is backcasting. Backcasting starts with a clearly defined vision of a desirable future based on normative criteria, then works backward to determine how that future might be achieved (Vergragt and Quist 2011). The gap between the envisioned future and the present is analyzed, identifying key obstacles, conditions, and potential leverage points for change. From there a set of strategic actions is developed to transition toward the preferred future. Backcasting encourages transformative thinking by breaking with dominant trajectories—in contrast to normative forecasting, which typically emphasizes incremental changes or continuity with the present. As such, it is particularly useful for addressing wicked environmental challenges that require systemic change rather than marginal improvements.

Another scenario-modeling approach is futurecasting, which extrapolates forward from current trends to explore possible outcomes. Often used to anticipate potential risks and opportunities, futurecasting begins by identifying key drivers of change (for example, climate trends, demographic shifts) and then analyzes how these factors might interact over time through multiple plausible future scenarios (Shamiyeh 2024). By mapping these trajectories, futurecasting helps decision makers prepare for uncertainty and design adaptive strategies.

In an era of biodiversity loss and socioecological transformation, these scenario-modeling methods provide structured ways to think critically about medium- and long-term change, helping to navigate complex future challenges. In this section, we describe a novel intervention that combines elements of both backcasting and futurecasting, integrating normative goal setting with exploratory and speculative scenario analyses. This hybrid approach maps a plausible, conceivable particular environmental future and charts a route toward the most desirable outcome.

The "Encountering Waste" intervention took place in a quiet, picturesque Icelandic valley, located at the gateway to the Westfjords and Snaefellsnes peninsula. The valley is known among locals as the home of Guðrún

Ósvífursdóttir, one of the standout characters in the *Laxdœla saga* (The Saga of the People of Laxárdalur) (ca. 1230–1260). On the morning of the second day of the walkshop, a dozen scholars of walking and environment gathered behind the Dalahótel, a family-run countryside establishment (and the only accommodation in the valley), awaiting their assignment. The task they were given involved a tactile encounter with something many would have preferred remain invisible: traces of human waste left behind by visitors to Sælingsdal, amid the area's expansive landscape of rolling grasslands, specked with green moss and lichen.

In the first component of the intervention, participants were instructed to walk a transect with a colleague, keeping an eye out for any instances of human waste on the ground. In ecology, a transect refers to a straight line or narrow section through a natural environment along which systematic observations or measurements are made. Transects are commonly used in environmental studies to assess biodiversity, monitor environmental changes, and examine the distribution of habitats and organisms across a specific area. For our intervention, we selected the hilly slope behind the Dalahótel as the site of the transects.

The participants were paired up and then sent off, with some choosing a relatively easy, unencumbered path while others bushwhacked their way up the gradual slope of the valley. The groups were instructed to call out "Stop!" whenever they encountered a piece of toilet waste or other types of trash. Each time a "Stop!" rang out across the valley, the entire line of walkers would come to a halt. Everyone would then wait in silence and anticipation for a few moments as one of the intervention's leaders recorded the artefact, before continuing along their transect. The walks lasted for twenty minutes, at which point the groups were instructed to reunite just behind the bathhouse at the geothermal spring. During this period, across roughly sixty square meters of land, the walkers recorded some sixty-seven distinct instances of waste, primarily consisting of bits of toilet paper and associated human fecal material.

Typically, when human waste is excreted in nature, microbial decomposition processes help break down the organic matter and recycle its nutrients (Ells and Lee 2000). Sunlight, soil microbes, and environmental exposure can facilitate this decomposition over time. Nitrogen and phosphorus are usually absorbed by plants or the soil. However, problems arise when decay cannot take place or occurs only at a slow pace—such as in northern latitudes or colder climes. Feces contain bacteria, viruses, and parasites that can pose health risks if not properly decomposed. While some pathogens die quickly, others can survive for days, weeks, or even longer, depending on environmental conditions. When waste enters a water source, it can introduce harmful bacteria, leading to waterborne diseases or algal blooms. Feces can also attract insects, scavengers, or other animals, which may spread pathogens from waste to other surfaces.

In the second part of our intervention, we invited participants to imagine a potential future for the landscape around them. The theoretical framework for this exercise was based on the "container model" of conservation and mobilities, a governance approach in which protected natural areas or national parks are treated as isolated "containers" of biodiversity or natural heritage, separate from the surrounding landscapes and human activities (Adams and Hutton 2007; Lapointe 2024). This model seeks to achieve conservation goals by establishing strict boundaries, limiting human activity, and prioritizing protection over sustainable use. We also drew on the concept of the acoustic soundscape in the exercise, which refers to the audible sounds produced by both humans and nonhumans in a given space.

Participants were split into three groups, each assigned a different method of engaging with and paying attention to the landscape around them: (1) walking a transect collectively while observing and scanning the ground, (2) walking a transect and paying attention to changes in the acoustic environment, and (3) walking a transect within a single acoustic environment, or "sound-sitting" within one place. Once these initial walks were completed, we invited participants to reenact the same walks, but this time imagining the landscape within a fictive future scenario. We told them that, at some point in the near future, a Saga era artefact buried in the Sælingsdal valley would be unearthed by archaeologists, generating a surge in interest in the region and contributing to an explosion in new nature tourism. This would result in the construction of a new canal and fleets of cruise ships navigating the interior of Iceland to unload thousands of tourists onto the doorstep of our quaint, family-run hotel. Participants were then sent off to rewalk their routes with this new context in mind. Once finished, we reconvened to discuss their reflections in the larger group.

In this intervention, the pauses that followed the incantations of "Stop!" became a collective reckoning for participants—a moment to reflect on what they were doing in one of Europe's most iconic landscapes, standing with their feet in, well, shit. These were sobering, humanizing moments. With attention focused on the visible traces of waste, the otherwise positive and pristine image of what nature could or should be seemed tarnished, marred, besmirched—colored, in some cases, from green and blue to brown and black. Participants reflected on the place of all this waste strewn across the earth and the role it might play in both our present and future. They thought biopolitically about how human nature is "governed" in nature, and how human waste and the accoutrements it accumulates—scraps of tissue, states of decay—are rendered into awkward states of being that must be managed.

Regarding the uninvited and often overlooked aspects of landscape waste, some participants expressed excitement as they focused on the task, their feet at the ready for encounters with untoward land. But their emotions shifted as they moved from start to finish along their transect. The initial encounters with waste were often giddy, almost gleeful. It was something new, unexpected, unknown. Participants expressed satisfaction at having recorded something that either should not have been there or was unanticipated in the landscape. Others expressed remorse, pointing to the contradiction of desiring freedom while feeling a need to control the behavior of others to prevent waste accumulation. Still others reflected on how much current social norms truly allowed for controlling the bodily behavior of others. The responses spoke to the micro-damage humans inflict on the imagination of pristine nature—one in which humans do not sully the environment, most certainly not with their own excrement.

The tarnishing of nature that could result from such a boom in tourism was not only visual; it was also palpably aural. Participants reflected on their experiences when unexpected sound "incursions" clashed with their notions of an acoustic environment indicative of wilderness or of a wild landscape, triggering a shift in feeling or tenor. These sound "wastes" most often came from the mechanized noises of cars, buses, or airplanes. The walkers discussed how the imagined presence of cruise ships and the onslaught of Saga-hungry crowds would produce an acoustic environment that would radically alter the sensory experience of visitors to Sælingsdal today. They were asked to consider this reterritorialization of place, to view the intervention as a conscientious practice of engaging with the undesirable in nature.

Transected Reflections: Behinds Leaving No Waste Behind

A key insight that emerged from this intervention was the role of biopolitics in shaping tourist behavior. Participants' reflections made it clear that social norms, policies, and conservation narratives play a crucial role in governing human-environment interactions. The individual and collective act of stopping, announcing, and recording an encounter with a piece of waste became an embodied form of environmental awareness—one that challenged the assumption that nature exists as a separate, untouched entity rather than a space continuously shaped by human activity. This suggests that addressing the problem of human waste in natural areas cannot simply be a matter of infrastructure development; it must also involve fostering critical reflection on the responsibilities of the visitors who temporarily populate these landscapes.

Participants reflected on the affective experience of walking a transect, with emotions ranging from disgust and sadness to excitement and encouragement. For example, when passersby expressed interest in the activity we were engaged in, participants were reminded that our rambling through the Icelandic hills actually had a real purpose. This prompted us to consider how our reflections might contribute to addressing the consequences of human (in)action. As wastes flow with people, they produce toxins and sounds. As people and their accoutrements become mobile, the notion of a "container" for protecting a given state of nature seems ill suited to the everyday realities of human-nature relations today. Instead actions aimed at conserving biodiversity and biocultural diversity must actively seek to foster and shape conscientious human waste practices.

The intervention also raises important questions about the future of conservation in high-traffic nature destinations. The picture we painted of a hypothetical mass tourism future in Sælingsdal served as a provocative lens for examining the limits of existing governance models. If visitation to this region continues to increase, as it has in other parts of Iceland (and the world), the dilemma of managing human waste will only become more pressing. However, rather than approaching this issue solely as a problem of regulation and enforcement, our intervention—walking transects of waste—points to the real potential of participatory methodologies to cultivate more engaged, accountable, and environmentally aware humans. By challenging the exclusionary "container model" of conservation, we advocate for a shift toward a more dynamic and relational approach—one that acknowledges the always, already entangled nature of humans and their environment and seeks solutions that are both ecologically and socially sustainable. This requires moving beyond merely a "Leave No Trace" ethic to address the systemic forces that generate overwaste in the first place. These forces range from tourism marketing strategies to chronic underinvestment in infrastructure.

The aim of our waste-walking intervention was to highlight the massive terraforming impacts of increased visitation in remote nature spaces without an affirmative biopolitics of tourism. The intervention underscored the constant demand on communities to become active participants in the appropriation of nature and on public authorities for appropriate, active stewardship that is at once ethical and sustainable.

Walking with waste may take different paths, as we have described in this chapter, but the trajectory from a single footstep to an overaccumulation of waste is the same. Our intervention highlights the importance of shifting the discourse on conservation from a punitive to an affirmative biopolitics—one that empowers individuals to see themselves not as passive consumers of nature but as active participants in, and producers of, its stewardship. Our hope is that by making waste visible and audible, and engaging visitors in collective reflection while walking with waste, we can help individuals and communities come to terms with our excrement and its aftereffects. Such interventions can reframe waste management not as an individual burden but as a shared responsibility. This shift in perspective is crucial for imagining conservation futures that do not merely seek to contain or erase human impact but instead foster deeper, more responsible engagement with the natural world.

As tourism continues to flourish in fragile ecosystems, which is only encouraged by many national economic goals, addressing the issue of human presence and waste will require more than just technical fixes. It will demand a fundamental rethinking of how we conceptualize nature, conservation, and the role of the visitor within these logistically (and epistemically) challenging landscapes. The waste walk is a compelling model for a participatory, embodied methodology that can contribute to this rethinking by providing both a critical lens on the present and a framework for imagining more sustainable tourism futures.

References

Adams, William M., and Jon Hutton. 2007. "People, Parks and Poverty: Political Ecology and Biodiversity Conservation." *Conservation and Society* 5 (2): 147–83.

Andrade, Riley, Carena J. van Riper, Devin J. Goodson, et al. 2023. "Values Shift in Response to Social Learning through Deliberation about Protected Areas." *Global Environmental Change* 78: 102630.

Ells, Michael D., and Kathryn J. Lee. 2000. "The Fate of Feces and Fecal Microorganisms in Human Waste Smeared on Rocks in a Temperate Forest Environment and Its Impacts on Public Health." National Outdoor Leadership School, National Park Service, Leave No Trace Center for Outdoor Ethics, Ferris State University.

Haklay, Mordechai, Daniel Dörler, Florian Heigl, Marina Manzoni, Susanne Hecker, and Katrin Vohland. 2021. "What Is Citizen Science? The Challenges of Definition." In *The Science of Citizen Science*, edited by Katrin Vohland et al. Springer.

Keller, Rose, Sofie Selvaag, Sigrid Engen, and Ole-Jakob Kvalshaug. 2025. "Testing Communication Strategies for Reducing Toilet Waste in a Norwegian Protected Area." *Journal of Interpretation Research* 19 (3): 10–16.

Korpilo, Silviya, Elina Nyberg, Kati Vierikko, Hanna Nieminen, Gustavo Arciniegas, and Christopher M. Raymond. 2023. "Developing a Multi-Sensory Public Participation GIS (MSPPGIS) Method for Integrating Landscape Values and Soundscapes of Urban Green Infrastructure." *Landscape and Urban Planning* 230: 104617.

Næringslivets Hovedorganisasjon (NHO). 2025. "Tall og fakta om norsk reiseliv." https://www.nhoreiseliv.no/tall-og-fakta/tall-og-fakta-om-norsk-reiseliv.

Lapointe, Dominic. 2024. "The Tourism Periphery: From Structural Hierarchies of Place to Relational Ontology." *Tourism Geographies*: 1–10.

Lapointe, Dominic, David Guimont, Alexis Guillemard, and Cassiopée Benjamin. 2021. "People, Place, Values: Living Lab as Social Innovation Processes for Tourism Communities." *Anais Brasileiros de Estudos Turísticos* 11: 1–8.

NCBI. 2021. "4. Faecal Pollution." In *Guidelines on Recreational Water Quality: Volume 1 Coastal and Fresh Waters [Internet]*. Geneva: World Health Organization. https://www.ncbi.nlm.nih.gov/books/NBK572627/.

Shamiyeh, Michael, ed. 2024. *Practices of Future-Casting: Ways of Sharing Imagined Tomorrows*. Birkhäuser.

Vergragt, Philip J., and Jaco Quist. 2011. "Backcasting for Sustainability: Introduction to the Special Issue." *Technological Forecasting and Social Change* 78 (5): 747–55.

World Travel and Tourism Council (WTTC). 2022. *Nature Positive Travel & Tourism: Travelling in Harmony with Nature*. https://wttc.org/Portals/0/Documents/Reports/2022/Nature-Positive-Travel-And-Tourism.pdf.

10

Folkloric Walking Methods for Environmental Entanglement

Tina Paphitis

Walking is integral to numerous vernacular customs and traditions, connecting participants to place and time through bodily motion and sensory engagement. These walking and walked practices can be as deliberate as representing or reenacting a narrative or event, or forming part of a processual or magical practice. They may be an incidental, functional necessity, or they may be metaphorical, reflecting the path of life in a ritual, tale, or song. Walking thus has a special place in folklore and folkloristics, both as a frequent theme or feature of customs and as a method of investigation—whether in antiquarian "walking tours," go-along interviews, or storytelling in motion. This chapter considers the possibilities of merging the two—by using a particular folk custom centered on walking as a research method, specifically aiming for environmental entanglement.

As Anna Tsing comments, "Walking is the speed for noticing—and for thinking" (in Shepherd et al. 2018, 1). I take Tsing's observation further, considering that arts of noticing might be cultivated not only through the *speed* of walking but also through *modes* of walking. I suggest that folklore can provide some of these modes. Looking at one example of folkloric walking, the *årsgång* (year walk), and its experimental application, this chapter explores how participants can form greater ecological entanglements by noticing and engaging with (real and imagined) more-than-humans, heightening senses of place, and disturbing time. In turn, I posit that environmental entanglement is a form of enchantment, mediated through a folkloric walk.

Walking, Nick Shepherd (2023) notes, can "flatten out hierarchies between theory and practice and between scholarly and creative practices" (59). This perspective is effectively illustrated here through engaging in folkloric walks—

which are vernacular, creative, and affective—as a research activity. The flattening of theory and method is also reflected in this chapter, where theory does not drive the method but is entangled with the act of walking. I mainly set out to conduct these experimental walks with a basic inkling: that we might be able to engage with environments in different ways through walks drawn from folk tradition. However, as an archaeology student, I had been drawn to Chris Tilley's (1994) and others' phenomenological approaches to landscape in archaeology. I carried these ideas into my folklore work (for example, Paphitis 2014), where they subconsciously underpinned much of my work in landscape and walking, and they help structure the experiential aspects of environmental engagement here.

Having undertaken methodological experiments in folkloric walking, beginning with a methodological legend trip (Paphitis, forthcoming) and, next, the *årsgång*, it struck me that the supernatural nature of many legends and oracular rituals that provoke such walks can give researchers some insight into the varied beings—"biological," "material," or "supernatural"—and the experiences of them that comprise the landscape, and what they might reveal about the ecologies of place. This, in turn, led to a consideration of these walks as more-than-human entanglements. This walk and my account of it are therefore entangled within the broad environmental humanities concept of the "arts of noticing" (Tsing 2015)—more-than-humans, temporalities, and multiple worlds. This is set within methodological frameworks that draw on some aspects of archaeological landscape phenomenology, although this theoretical perspective is seen here as supportive rather than instructive.

My use of these concepts grew from the humbly observed seed that folklore has long explored relationships between humans, nonhumans, things, and environments. The popular association of folklore with nature unfortunately stems largely from persistent, erroneous (and dangerous) notions that later folklore was a remnant of nature worship and fertility cults (for example, Frazer 1890; 1926), as well as from the "naturalization" of peoples with a particular land through their folklore, as implied in romantic nationalism (Abrahams 1993). However, academic studies and emic reflections support the strong connection between folklore and the environment in diverse ways (for example, Honko and Löfgren 1981; Abram 2019; Vickery 2019; McDowell et al. 2021). In doing the walk, and reflecting extensively on it, on my colleagues' responses, and on the nature of folklore, I came to the realization that environmental entanglement is a form of enchantment, with these walks serving as the enchanting act. Rather than pretending that this is a theorization I set out to test from the start, I will explore it as part of the discussion on the experimental walk at the end.

Experiential Walking for Environmental Entanglement

The typical study of folkloric walks is to examine the performance of the walks themselves, and their performers, in the contexts in which they are performed. This is not, strictly speaking, what is being studied here, except insofar as it provides background to the walk and its regular contexts. The primary purpose of undertaking the walk here is as an experimental—and experiential—research method, aimed at assessing how these walks might impact or enhance our engagement with the environment. I consider this a reflective walk, exploring both my own and my colleagues' entanglement with the ecology of place. It also extends or reenlivens phenomenological approaches to landscape, outside of—or in addition to—environmental philosophy's contribution to the environmental humanities.

Rather than employing phenomenology as part of environmental philosophy, I build on the basics of archaeological methodologies for a phenomenology of landscape (Tilley 1994; Hamilton et al. 2006), which primarily draws on the phenomenological formulations of Heidegger (1962 [1927]) and Merleau-Ponty (2002 [1945]). Especially pertinent to the environmental humanities, phenomenological approaches emphasize the dissolution of nature/culture and mind/body dualities, further underscoring the methodological approaches to walking noted earlier. At the core of archaeological landscape phenomenology is the examination of experience with and in the world. The concept of being-in-the-world is central, where humans engage in intense dialogic relationships with the landscape and what it comprises: its paths, places, monuments, and other Beings. Such an approach is thus specifically grounded in relationality between people and landscape. Landscapes in archaeological phenomenology are social. However, while archaeological landscape phenomenology makes much of the relationality between people, paths, things (monuments), geological features, seasons, weather, and trees, its attention to the varieties of the more-than-human and their affects is somewhat limited. This is expanded here to include the relationality of how all aspects of the environment are entangled and affective—sociality is found in the nonhuman as well as the human. Here the human researcher is experientially immersed in such environmental entanglements through the active application of a folkloric walk.

Archaeological landscape phenomenology emphasizes both the experimental and the experiential. Although it has been criticized as methodologically weak due to its subjectivity and particularity (see Hamilton et al. 2006, 34, for a summary), it is these very qualities that are of particular significance here. Every engagement with the environment is specific, new, and grounded in the particularities of place, time, and sensory experience. Walking is key to phenomenological methods, with an emphasis on the sensory. While sight is the most commonly used and discussed, all senses are significant. This is also the case for the *årsgång*: as we will see, to prepare for the *årsgång*, practitioners must undergo a process of sensory (and social) deprivation to heighten their senses during the walk. This approach was applied as fully as possible in this walk.

There are notable connections between phenomenological or experiential approaches to landscape and the methodological approach of the "arts of noticing." The concept of the arts of noticing has been developed and applied within the environmental humanities as a means of enhancing awareness of, and engagement with, nonhumans and the more-than-human world. Arts of noticing encourage attention to small, unnoticed, or overlooked details of the world, cultivating a deeper understanding and entanglement with our environments (Tsing 2015). Through this practice, we can better appreciate the sometimes fragile interweaving of human-nonhuman relationships. Practices that foster arts of noticing can lead to unexpected encounters, valuing the sometimes unpredictable ways in which we engage with the world and what we find in it. Fundamentally, then, arts of noticing are about experience and encounter, about the particular and the marvelous, and about connections and relationality—much like phenomenological approaches to landscape. The experimental *årsgång*, and this chapter, therefore, seek to develop a specific walking methodology that applies the concept of the arts of noticing, drawing on earlier methodological developments in archaeological landscape phenomenology.

Using approaches in landscape phenomenology and the arts of noticing to engage with the environment as a basis for my walking practices, I introduce a third dimension: the specific mode of walking. This mode, drawn from folk tradition, provides methodological structures for experiential, experimental, and more-than-human encounters with the environment. The *årsgång* was selected here precisely for its attention to time, space, the potential for more-than-human encounters, and, especially, its embodiment/sensory aspects. It aims at methodological innovation through engagement with traditional practice, while also emphasizing the intimate socioecological relationships fostered by folk groups.

The connection between walking and folklore as a research method is not without precedent. Matthias Egeler (2021) has considered how, by traversing the landscape on foot, one can better understand the legends connected to it, and their everyday contexts, through what he calls "close walking." I suggest that this approach can also be extended to "reading" or immersing oneself in environments, both legendary and otherwise. In other cases, sensory walking that emphasizes sound and performativity has been used "to excavate, actualize and interpret" folklore associated with particular landscapes (Anemogiannis and Theocharis 2021). Rather than using walking through environments as a way to better understand specific folklore, I use folklore-as-walking-practice to gain a deeper understanding of environments.

Årsgång

Much of the information we have regarding the Swedish *årsgång* (year walk) has been skillfully synthesized by Tommy Kuusela (2016), and the summary of the *årsgång* as a traditional practice here is drawn from his work. As Kuusela explains, the *årsgång* was a "special technique of divination" (2016, 65), where the walk, combined with specific preparations and spatial and temporal conditions, was performed to foresee what the coming year would bring—who would die, who would marry, whether harvests would be good or fail, or any other significant events.

The *årsgång* is tied to specific locations in time and space. It is a custom that is no longer practiced today (or, at least, not recorded as being practiced, with any contemporary performances likely being revivals rather than continuations).[1] It is documented from the seventeenth to the mid-twentieth centuries. Records indicate that it was typically performed on Midsummer's Eve, Christmas Eve, or New Year's Eve—special times of the year when boundaries between human and supernatural worlds are thinned or dissolved, or held as particularly magical. The time of day was also important: the walk was usually performed at midnight or before dawn.

Spatially, the *årsgång* is principally documented in southern Sweden, though there are several examples from other parts of Sweden and a few from the rest of the Nordic region. It was predominantly practiced in rural communities, where participants often moved away from (living) human dwellings to walk through meadows, fields, and forests, as well as on crossroads and church paths. The church was a common focus of many year walks: by walking the paths leading to the church, the year walker could glimpse the funeral processions of the coming year, thereby acquiring knowledge of who would die. In other instances, this was achieved by circling (*kringgång* in Swedish—"round walk") the church, sometimes widdershins (counterclockwise), sometimes three times, and then breathing into or looking through the keyhole of the church door to witness future events within. Other buildings, such as homes, might be encircled to produce visions. Walking in or by fields could reveal the outcome of the coming year's harvest, or the year walker might see fires, acts of violence, or their loved ones.

The *årsgång* is a solitary activity. Year walkers must set out and complete the walk alone. They must not speak to anyone the entire evening beforehand, nor make a sound while walking—including when encountering visions, otherwise they would disappear, and no further omens would appear to the walker. Supernatural beings might sometimes appear on the year walker's path, attempting to break their concentration, whether by crying out, responding to them, laughing at their antics, or otherwise causing them to lose focus on their task. Focus is achieved through other preparations made in advance of the *årsgång*: in addition to refraining from speaking, the year walker must not eat or drink anything the entire evening beforehand, nor must they look into a fire.

Going on a year walk could thus be dangerous—physically, due to the preparations the year walker had to make and the isolating paths they tread, but also mentally, emotionally, and spiritually, because of the uncertainty and the potentially distressing sights or harmful beings they might encounter. However, those who completed the walk successfully were not only rewarded with visions and knowledge of the coming year, but their powers of sight could also grow with each year, with some allegedly gaining permanent second sight or other extraordinary abilities.

Undertaking a Year Walk

The experimental year walk was part of a session within a One by Walking workshop held during the daylight hours in Sælingsdalur, Iceland, in August 2023, disrupting its spatial and temporal aspects. The goal of the session, cofacilitated by Simon Poole and me, was to explore the bodily and rhythmic meanings of ritual walking and how they can be reconfigured to offer new ways of understanding the environment, time, and self—through contextual mistimings, spatial relocations, and differential social groups. After introducing the overall aims of the session, we presented two folkloric walks that the group, divided in two, would attempt. With one half of the group, Simon led an activity inspired by the Royal May Day procession from Knutsford, United Kingdom, while I led the *årsgång* activity with the other half. The two walks were intentionally different: the May Day walk is undertaken in a group, loud, and full of flamboyant movements, while the *årsgång* is somber, individual, and discreet. We asked the participants to bear the following questions in mind while undertaking their assigned walks:

- Which senses feel heightened, deprived, actively engaged, or suppressed?
- What do you notice about your own rhythms and the rhythms of the environment?
- What sense of community—human and nonhuman—do you feel (or not)?

I described the tradition of the *årsgång*: its contexts, purpose, and method. Because the activity was held at four o'clock in the afternoon, and we had made the long drive from Reykjavík starting early that morning, I could not deny the year walkers food or drink beforehand, but I instructed them to refrain from eating or drinking during their walk. They were also instructed to walk completely alone, in total silence, with their environmental engagement and the purpose of the tradition in mind. They were asked to not use their mobile phones – a kind of "false fire," and a distraction. We set off on our walks for half an hour, after which the group reconvened to discuss and compare their experiences.

We walked a short distance from our own "settlement," the Dalahótel. The area featured diverse vegetation—mainly bushes, low trees, grasses, and mosses—along with streams, a thermal pool, and structures, including a bridge and hut (the latter serving as a changing space for pool users). All the year walkers adhered to the instructions of silence, fasting, and solo walking. Whenever our paths crossed, we did not acknowledge the other walker through looks, gestures, or speech but gently altered our paths to avoid them. In this way, we maintained our focus on our walks. The restrictions of the *årsgång* induced us to concentrate more strictly on our paths and our environment—that is, they cultivated our arts of noticing. In the isolation of our social thoughts and human bodies, we were able to more closely attend to the nonhumans we encountered.

Figure 10.1 Sælingsdalur *årsgång* (images taken after the walk so as not to impede it). Source: Photo by Tina Paphitis.

It took a moment or two to focus my attention on the walk, but it was the act of walking itself that helped me do so. Looking down, step by step, I could feel the connection between the landscape and myself deepen through my feet. I sensed that the long grass beneath me concealed all manner of communities, and I was conscious of where and how I trod. My social connections with my colleagues faded into the background as I cultivated the society of Sælingsdal's nonhumans. In consciously walking away from our group's meeting place, the hotel, and the May Day revelers, I entered into a deeper immersion with the place and its denizens.

Instinctively, I walked over to a group of trees: mainly a conference of poplars and dancing, downy birches. The poplars felt protective, sheltering me from the breeze, and their solid rootedness and tight branches contrasted with the darting of a bird in and out of their depths. I wondered how long the bird and its family had resided there, and how long they could remain undisturbed. I had a sudden concern that someone might come along and disturb them, but that thought flew by just as quickly as it had come. The birds are also their own agents and know better what it was to dwell in these trees than I do. The poplars, too, were better protectors of the birds than I could be. I stepped closer to the companion birches. Their movement and the intertwining of leaves and stems seemed all encompassing. I have an innate aversion to trees being cut, and I flinched at the thought that these playful branches might be curtailed in the name of "landscape management"—or in an even more aggressive clearance of the land. As I walked closer to them, though, their liveliness encouraged me. I could almost see—and almost hear—their stretching and growth, which would continue long after I had passed them.

Arching my course around the trees, I wended my way to the small bridge spanning a branch of water running through that part of the dale. The bridge invited me across, and the hollow thud of my steps on the tarmacadam-decked wood jarred in comparison to the bumpy, tangly long grass. On reaching the other side, I noticed the grass was shorter and petered out into gravel. A parked car stood as a sentinel, indicating the presence of other humans, nowhere to be seen. I felt I was on the edge of one realm and another, and it was unsettling. I didn't want to be in this new place—it didn't speak to me, and I felt exposed in it. I turned back and walked along the water, reimmersed in the place, which was showing me its ways of life. Surveying the deep banks along the water, I found a place to leap across.

I made a deliberate line for the other, more open stream, crisscrossing it a few times, stepping over stones in the shallower or dried up parts. The stones muttered and grumbled underfoot, but I thanked them for supporting my steps, if somewhat precariously. Walking beside the stream, against the current, I felt slowed and dragged by it. I stopped and listened to its song, then sat among the blonde and verdant grasses and felt the water. The shock of its heat, for the waters there are thermal, was startling. Yet it was comforting. It spoke of the connections between underground and overground, grounding me in that particular landscape. Ever running, the water, too, flowed into the future. I saw a time when its course might be altered, or dry up, or choked by filth of heedless visitors. The vision was disturbing but soon dissipated. I felt the presence of the *huldufólk*, Iceland's "hidden people," most strongly at this point. Their *álfakirkjan* (elf-church) of Tungustapi stood approximately a kilometer away (Figure 10.2), and this landscape was just as much theirs as it was that of their "scientifically real" nonhuman neighbors and companions. Their presence contributed to the sacrality, otherworldliness, and liveliness of the Sælingsdal landscape. They were evidence of strong human-environmental entanglements, a reminder of the ongoing respect for the environment, their home. I looked down again. In the troubled water, I saw the future of a situation of my own, joyous for now, but that would move to inevitable sorrow.[2] I silently blessed the waters, then moved on. I was concerned for all the seen and unseen things, whose labor had formed this place, enabled it to thrive, and granted me leave to commune with it. But these things are hardier than I, and their collaboration may long endure after I had gone.

The group reconvened, and we broke our silence. Some seemed to take a moment to return to the present and to our community of human colleagues. But many smiled as we gathered and could not wait for the carousing processioners to regather and refocus, so that we could share our impressions of the walk.

Figure 10.2 Tungustapi, Sælingsdal. Photo by Tina Paphitis.

"I enjoyed that" was the first comment, followed by agreements from others.

"Yes, I got really into it" came the next. "It went really quickly. I didn't want to stop."

We discussed the walk in general, as well as in relation to the specific questions noted earlier. The year walkers felt that all their senses were heightened, especially those other than sight. One said that the walk drew her attention to the darkness and the light. The focused nature of the *årsgång* required an intensity of engagement that resulted in increased stimulation. This was likely further enhanced by the explicit goal of this particular year walk: to utilize the bodily and rhythmic aspects of ritual walking in order to immerse oneself more deeply in the environment and consider both the present and future through the lens of a reconfigured past custom. The walk helped the walkers *notice* more, fostering an intimate relation with their surroundings. Their rhythms and steps slowed, and they became much more attentive to their own bodies and ways of walking, synchronizing them to the environment. Both the slowness and dynamism of the environment were infused into the walk and the walkers. This gave them the space and time for their minds to work in harmony with their bodies, at the same pace. The physical act of walking impacted their mental perambulations, underscoring the idea, introduced at the beginning of this chapter, that walking brings together mind and body—or, more accurately, dissolves such dualities, as in phenomenological perspectives.

Despite the recontextualization of place and time (both the year and the day), all the year walkers remarked that the *årsgång* felt fitting for the Icelandic environment. It was an apt way to engage with the landscape, and with time, through bodily rhythms and restrictions. One year walker mentioned having walked around the hut three times, feeling as though she was performing a magical action. The recurring steps on the same ground created, with each repetition, an increasing sense of familiarity. The circling motion at once disorientates and grounds the walker. Significantly, the walkers reported feeling closer to the environment and its diverse inhabitants. While no one explicitly mentioned supernatural beings,[3] some responses hinted at something marvelous and spellbinding evoked by the walk.

Entanglement as Enchantment

In undertaking this walk and analyzing its effect on myself and my colleagues, it became clear that such walking practices can help participants—including researchers—entangle themselves more deeply with environments. While this has long been the purpose of many traditional and ritual walks, it has rarely, if ever, been considered a tool in academic research. Such walks encourage participants to pay attention, to notice. The year walkers here reported a deep immersion in the walk—and consequently, in the environment: "got really into it" was an expression of this entanglement. Thus, it was the mode of walking that facilitated the arts of noticing. One could center the discussion of this experimental year walk solely on the notion of environmental entanglement. Yet in reflecting more closely on this activity, and its contextualization within broader studies in folkloristics, it occurred to me that what the year walkers, myself included, encountered was not just entanglement. "Got really into it" is also a statement of enchantment.

The concept of "enchantment" has been used in numerous disciplines, particularly in the field of environmental humanities,[4] to varying purposes and effects. Yet enchantment is, first and foremost, found in vernacular encounters with places and more-than-humans: that is, in folklore, with folklore studies best placed to understand the phenomenon. Folkloristic work on enchantment is extensive, and a comprehensive summary and assessment is not possible here. The essentials of enchantment here are mainly drawn from work by Asplund Ingemark (2006), who, appropriately, has examined the multispatiality and multitemporality of enchantment as seen in folk narratives, through an analysis applying Bakhtin's (1986) formulation of the "chronotope."

In folk narratives, enchantment manipulates protagonists mentally, physically, and emotionally, affecting their perception, use of their body, senses, notions of time, and personality (Asplund Ingemark 2006, 1). We can see that the preparation and performance of the *årsgång* facilitate—or affect—this manipulation of the body, mind, and senses, first through social and sensory deprivation, and then through heightened awareness and engagement with the environment. This sensory deprivation, in preparation for sensory enhancement during the walk, allows the walker to become more alive to their being-in-the-world—and to *beings*-in-the-world—encountering place and Beings in ways that might otherwise not occur in everyday life.

Enchantment can "transform a well-known human landscape into alien territory in an instant, introducing elements that simply do not exist in the normal, everyday world" (Asplund Ingemark 2006, 9). This "territory" can manifest on the spatial level—revealing hidden worlds with unusual denizens—and on the temporal level, as shifts in time. These multiple worlds can be seen in the experimental *årsgång*—from the ability to "see" microscopic or concealed elements of the environment, to witnessing "future versions" of nonhumans or landscapes, to sensing the presence of supernatural beings. Matter is activated and affects the human wanderer: the inviting bridge, the cautionary car, the responsive stones. The sense that I was walking from one space into another felt very real, though otherwise the transition would barely be noted. The multitemporality of the *årsgång* is evident in the shifting between a past custom, grounded in the present, and a glimpse into the future. When enchantment takes effect, it is expressed in the futurity of what walkers witness—or seek to understand—and in the unusual passage of time: it went quickly for some walkers, while for others it took a few moments to refocus on the present and recall their presence—encompassing both differential space and time.

Such encounters are not normally possible, but the year walk makes them possible. It also allows researchers to *be* enchanted, to explore the possibilities and multiple worlds and temporalities of environments. This is not about "belief": I do not believe in supernatural beings, but the *huldufólk* were present in the landscape due to the enculturation of the "natural" world into these beings, and the numinous quality of the landscape imbued by the walk. In folkloristics, the term "ostension" is used to describe behavior that refers to or emulates something legendary (see Dégh and Vázsonyi 1983). Researchers, too, can engage in ostensive practice—"be enchanted," here—and use it as a tool for research. As environmental humanists, we often advocate for engaging with the creative, affective, and enchanting; this is one way to do so.

As a traditional practice, the *årsgång* is clearly an act of enchantment. It is a means by which the year walker can access multiple worlds (including those inhabited by ghosts and other supernatural beings) and multiple times (the future) through an alteration of the senses. Distraction, or taboo action, can break the enchantment. Similarly, the experimental year walk conducted here is arguably an enchanting act, designed to better entangle researchers in the environment. The walkers sensed things they would not normally perceive and were immersed in a place and time quite apart from the "normal" world. Like the distractions affecting the traditional year walker, my enchantment broke down when I began to think about another, personal, issue rather than maintain focus on the task—or be distracted or disturbed by what the enchantment was showing me. Enchantment is not always benign or positive: it can be dangerous or distressing, underscoring the complexities of engaging with the environment, which can be both despairing and hopeful. The year walk can form and reveal these complex entanglements in unique and affecting ways. It is often the case that the enchanted person is never the same again, and the hope is that by undertaking this experiment, the year walkers have experienced some shift in their perspectives or approaches.

Conclusion: Walking Traditions as Method

In this summary of and reflection on a single experimental walk, I have attempted to show how we might draw on folk customs for methodological inspiration. My approaches to landscape engagement in folklore have been grounded in phenomenological approaches in archaeology, which foreground the experiential, sensory, and relational aspects echoed in the environmental humanities' arts of noticing. The development of specific modes of walking, attuned to the speed of the walk, can provide systematic approaches to cultivating the arts of noticing while also allowing for the subjectivity and specificity of each activity. The *årsgång*, or year walk, offers some methodological guidance for walking as a means of cultivating attentiveness. The use of the *årsgång* enhances engagement and entanglement with place and landscape, engendering an art of attentiveness that draws the researcher into its ecologies. The experiential focus of the *årsgång* directs walkers' attention to the nonhumans of Sælingsdal and conjures future versions or supernatural beings.

This experimental walk shows the potential for drawing on various folk customs to develop walking methods in environmental exploration. We must, of course, be mindful of the respectful use of tradition, ensuring we do not inappropriately claim the traditions of others. Instead we draw on our understanding of customs and their purposes to inspire our research. Given the removal of various factors that make the *årsgång* a folkloric practice, we as researchers are not engaging in a folkloric tradition per se but are rather using these practices to explore new lenses and ways of knowing and noticing. We employ a folkloric walk as an act of discovery, ritualizing research-based walking. This ritualization provides a processual and interpretive approach to walking while freeing researchers to engage with the more imaginative, creative, and enchanting aspects of environmental entanglement. This experimental *årsgång* has demonstrated that we can employ enchantment as a method (cf. Stainova 2019) to effectively and affectively explore environments.

We could repeat the experimental *årsgång* in new places, at different times, and with various forms of embodiment. While this walk was undertaken in a remote location, it could easily be extended to urban environments. We might also apply the rules of the customary *årsgång* more stringently, abstaining from food, drink, light, and social interaction as traditionally prescribed. Given its subjectivity, participants should write their own accounts of the walk to better reflect their own voices and the myriad voices they are entangled with. One advantage of the *årsgång* is that it is conducted by a solo walker, giving the researcher the freedom to undertake it outside the constraints of specific research structures (and funding). However, as we have seen, it can also bring solo researchers together, contributing to a polyphony of environmental exploration.

Notes

1 In 2013, Swedish developer Simogo released *Year Walk*, a single-player adventure game for iOS (subsequent versions were released for other platforms), based on the tradition (see Vahlo 2018, 9; Höglund 2020, 182–84).

2 This did, indeed, come to pass.

3 This might have been due to a fear of sounding strange, or because such beings were simply not part of their experience.

4 In recent years, it has become popular to speak of "disenchantment" and "reenchantment" in the environmental humanities (see, for example, Murai and Parham 2021). I reject this Weberian assessment of so-called modernity as disenchanted, and the idea that we are now in a process of reenchantment. Note that the overwhelming majority of what folklorists study—much of which is enchanting in some way—is practiced within the thick of the capitalist, industrialist, and (post) modernist project, including the traditional *årsgång*. I consider "misenchantment"—misplaced enchantment—to be a more apt term to refer to the distracting forces of the market, which aligns with Radulović and Belić's (2021, 17) observations that aspects of the modern world "obstruct" magical, enchanting forces. The *årsgång*, as a practice, removes them.

References

Abrahams, Roger D. 1993. "Phantoms of Romantic Nationalism in Folkloristics." *Journal of American Folklore* 106: 3–37.

Abram, Christopher. 2019. *Evergreen Ash: Ecology and Catastrophe in Old Norse Myth and Literature*. University of Virginia Press.

Anemogiannis, Dionysis, and Angelos Theocharis. 2021. "'Echoes of Elves and Demons in the Riverbank': The Exploration of History and Folklore in Kythera through Walking." In *Experiential Walks for Urban Design: Revealing, Representing, and*

Activating the Sensory Environment, edited by Barbara E. A. Piga, Daniel Siret, and Jean-Paul Thibaud. Springer.

Asplund Ingemark, Camilla. 2006. "The Chronotope of Enchantment." *Journal of Folklore Research* 43 (1): 1–30.

Bakhtin, Mikhail. 1986. *The Dialogic Imagination: Four Essays by M. M. Bakhtin*. Edited and translated by M. Holquist. University of Texas Press.

Dégh, Linda, and Andrew Vázsonyi. 1983. "Does the Word 'Dog' Bite? Ostensive Action: A Means of Legend Telling." *Journal of Folklore Research* 20: 5–34.

Egeler, Matthias. 2021. "Close Walking: Musings from Goat Mountain." *FF Network Newsletter* 55: 5–10.

Frazer, James George. 1890. *The Golden Bough: A Study in Comparative Religion, in Two Volumes*. Macmillan and Co.

Frazer, James George. 1926. *The Worship of Nature, Volume I*. Macmillan and Co.

Hamilton, Sue, and Ruth Whitehouse, with Keri Brown, Pamela Coombes, Edward Herring, and Mike Seager Thomas. 2006. "Phenomenology in Practice: Towards a Methodology for a 'Subjective' Approach." *European Journal of Archaeology* 9 (1): 31–71.

Heidegger, Martin. (1927) 1962. *Being and Time*. Translated by John Macquarrie and Edward Robinson. Blackwell.

Höglund, Johan. 2020. "Nordic Gothic New Media." In *Nordic Gothic*, edited by Maria Holmgren Troy, Johan Höglund, Yvonne Leffler, and Sofia Wijkmark. Manchester University Press.

Honko, Lauri, and Orvar Löfgren, eds. 1981. *Tradition och Miljö: Et Kulturekologiskt Perspektiv*. LiberLäromedel.

Kuusela, Tommy. 2016. "'He Met His Own Funeral Procession': The Year Walk Ritual in Swedish Tradition." In *Folk Belief and Traditions of the Supernatural*, edited by Tommy Kuusela and Giusepe Maiello. Beewolf Press.

McDowell, John Holmes, Katherine Borland, Rebecca Dirksen, and Sue Tuohy, eds. 2021. *Performing Environmentalisms: Expressive Culture and Ecological Change*. University of Illinois Press.

Merleau-Ponty, Maurice. (1945) 2002. *Phenomenology of Perception*. Translated by Donald A. Landes. Routledge.

Murai, Mayako, and John Parham. 2021. "Introduction: Enchanted Environments." *Green Letters* 25 (2): 107–14.

Paphitis, Tina. 2014. *The Place of Folklore in Archaeological Landscapes: Narratives and Identity in Medieval to Modern Britain*. Doctoral thesis, University College London.

Paphitis, Tina. Forthcoming. "Walking with Ghosts: Legend Tripping as Method." In *Walking in Folklore and Ethnology*, edited by Tina Paphitis, Susanne Österlund-Pötzsch, and Simon Poole.

Radulović, Nemanja, and Smiljana Đorđević Belić. 2021. "Disenchantment, Re-Enchantment and Folklore Genres: An Introduction." In *Disenchantment,*

Re-Enchantment and Folklore Genres, edited by Nemanja Radulović and Smiljana Đorđević Belić. Belgrade Institute for Literature and Arts.

Shepherd, Nick. 2023. "Walking as Embodied Research: Coloniality, Climate Change, and the 'Arts of Noticing.'" *Science, Technology & Society* 28 (1): 58–67.

Shepherd, Nick, Christian Ernsten, and Dirk-Jan Visser. 2018. *The Walking Seminar: Embodied Research in Emergent Anthropocene Landscapes*. Reinwardt Academy Press.

Stainova, Yana. 2019. "Enchantment as Method." *Anthropology and Humanism* 44: 214–30.

Tilley, Christopher. 1994. *A Phenomenology of Landscape: Places, Paths and Monuments*. Berg.

Tsing, Anna. 2015. *The Mushroom at the End of the World: On the Possibility of Life in Capitalist Ruins*. Princeton University Press.

Vahlo, Jukka. 2018. "Video Game Studies and Contemporary Folkloristics." *Folklore Fellows Newsletter* 52: 5–14.

Vickery, Roy. 2019. *Vickery's Folk Flora: An A–Z of the Folklore and Uses of British and Irish Plants*. Weidenfeld & Nicolson.

11

And That Has Made All the Difference

Two Marches from Tel Aviv to Jerusalem

Dani Schrire

On a cloudy November day, I left Tel Aviv with a few hundred marchers. The rainstorm that caught us on the road on the outskirts of Tel Aviv left us soaked. I walked with a colleague I had bumped into, and we shared our thoughts. We were joined by a woman about my age, a psychologist who turned out to be a wonderful companion. I asked her if she had participated in the political march from Tel Aviv to Jerusalem four months earlier, in July. She was somewhat puzzled by my question and responded with her own: “Is there anybody here who did not participate in the July march?” Of course, this was an exaggeration. Yet here I was, marching again along the same route, again with thousands of people, many of whom were again the same activists, again confronting our government. And yet I cannot think of two more different marches.

The two marches discussed here took place four months apart—one in July 2023 and the other in November 2023—following a similar route that began in Tel Aviv and ended at the Israeli Parliament in Jerusalem (about sixty kilometers). The first march drew, at its peak, around seventy thousand to one hundred thousand marchers on the final stretch to Jerusalem, while the second had about twenty thousand participants. Like myself and this random companion who joined me on the first day of the November march, many participants took part in both, sharing a sense of repetition. Yet it took a while to recognize that, despite involving the same marchers on a similar route, the two marches had almost nothing in common. The gait, rhythm, body posture, tempo, feelings and emotions, words spoken, and much that was unsaid drew on a completely different walking repertoire and distinct marching histories. This chapter begins with an account of these two marches, paying close attention to body techniques, affects, sensorial aspects, and emotions. I then attempt to integrate the marching practices that characterize these marches into a broader discussion of how marches draw on a complex history and walking know-how, and how body memory is used here for comparison.

A brief political and historical context is necessary. I have been marching intensively as part of a very broad movement, with hundreds of affiliated organizations that have different, often conflicting agendas but are united in their opposition to Israeli prime minister Benjamin Netanyahu and his government. The protest movement against Netanyahu first emerged in connection to serious corruption allegations against him. The movement gained momentum during the first year of the COVID-19 pandemic, with many marches taking place, most of them in Jerusalem. Although the marches varied greatly week to week, a collective marching language began to form within this protest movement.

This movement reincarnated after the Israeli government's push for judicial reform, which was viewed as a judicial overhaul by most experts of constitutional law and considered closer to a coup by the protesters (Medina and Bloch 2023). For nine months, starting in January 2023, protests became extremely intense, and some weeks I would go out several times for hours on end. It is estimated that around 20 to 25 percent of all Israelis participated to some degree during that time, attending at least one demonstration or march. Hundreds of thousands took to the streets every week, in a determined, forceful, and nonviolent movement (Medina 2023). This momentum was brought to a complete halt on October 7, 2023, following the Hamas attack in the southwest of Israel. It took about a month for people to gradually return to the streets, initially to support the families of the Israeli kidnappees in Gaza. In the months after the attack, it became clear that the Israeli government, particularly Netanyahu, was not invested in securing the kidnappees' release and seemed more interested in revenge and maintaining its hold on power.

The two marches discussed here represent two distinct phases of this protest movement. The July 2023 march was part of the struggle against the judicial overhaul, while the November 2023 march focused on pressuring the Israeli government to negotiate an exchange that would secure the return of the kidnappees and ultimately end the current war in Gaza.

Studying Political Marches: Methodological Considerations

During these four intensive years of protesting and the two marches of the past year, I had no initial plan to research this movement. Only after the November march did it become clear to me that I wanted to study it. Like many fellow protesters who took photographs and video recordings, I documented the protests, though I preferred capturing the situations unfolding around me through audio recordings rather than visual documentation. I thought that if I ever studied these protests—or wished to reflect on these experiences—it would be easy to find photographic documentation. Sound, however, felt more evocative and would allow me to reflect on intensities, atmospheres, feelings, imaginations, and affects—all things that cannot be fully captured in words or images. My body became attuned to feeling, sensing, and immersing itself in what was happening, and audio recordings became a way for me to grasp the development of my protesting techniques and habits over these years.

Language is limited when it comes to representing the intensity of demonstrations—the crowds, the way your shoulders rub up against others, the smell of sweat, the cheers that get stuck in your mind for hours, the noise that stays in your body, police horses marching toward you, the cold water sprayed in the winter, or the stinky water blasted by the police, with a revolting scent that lingers for days. Academic articles fail to convey these sensory experiences, as well as the feelings of rage, fear, pain, and exhilaration. This gap exists not only in writing but also in the documentation process—photographs, sound recordings, and videos fail to capture the full sensorial and emotional overflow. Because recording devices always normalize sound, it too loses this magnitude. Comments such as, "You cannot imagine what I just experienced," "There's no way you can understand," and "Those who weren't there will never grasp this experience" point to the limitations of documentation and communication, which are seldom so clear-cut.

Methodologically, I use my own sound recordings, as well as available photographic documentation, but I analyze them as much as possible with attention to my bodily memory. My informal talks and ethnographic interviews with members of this protest movement helped me transcend my own bodily experience and move toward something closer to an intersubjective analysis of demonstrative events. Because my body relates to these two marches—and, in fact, to many more (fifty to a hundred in recent years)—body-memory becomes a vehicle for comparison between them.

A final word on ethics. My work has not been based on participatory observation. I did not march or demonstrate with the primary intention of observing and then writing about it. Given the importance I attribute to the interplay of observation and bodily performance, it is crucial to stress that I am part of a protest movement and am reflecting on it through an autoethnographic lens, combined with media analysis and ethnographic interviews. The clearest commitment I made from the start was that everything I do during a protest—where I stand, how I march, what I say, and what I chant—is guided by my sense of what will be most effective for the success of the protest. This did not change when I realized I was investigating these protests. My primary ethical commitment in participating takes precedence over my goal of documenting the protest or finding the best angle to observe it from in order to "tell the best story." In such public events, with hundreds of cameras around, the idea of consent becomes less relevant. What seems more important to me is the trust that I am a genuine protester, a real participant, not someone conducting research. Whenever possible, I prioritize bodily presence over the capacity to observe and represent. Dialogic reflections in interviews are, of course, very different; but even then, I engage in them as a fellow protester, acknowledging my commitment to the movement and the biases this entails as I try to learn about it.

July 2023: Marching as a Reenactment of Exodus

By July 2023, the protests against the judicial coup pushed forward by Netanyahu's government were in full swing. The government's actions and the protest movement were focused on a critical step in this process: a law brought to parliament that would cancel the rule of reasonableness. This rule, though rarely used, allowed the judiciary to annul legislation that lacked logical justification, thus shielding citizens from the unchecked use of power (Kremnitzer 2023). On July 18, another day of disruption began at 6:00 a.m. By nightfall, the Ayalon Highway—the main route running through the Tel Aviv metropolitan area from north to south—was blocked by one thousand marching protesters. Clashes with the police were inevitable; water cannons were used, and many protesters were arrested.

In parallel, a group of protest leaders announced in the early evening that at 8:00 p.m., they would march to Jerusalem. A hundred marchers set off, camping near Tel Aviv. The next day, the number of marchers grew to about one thousand during the day, and doubled by nightfall, when they camped near the airport. I joined the marchers on the third day at their long stopover for lunch in a shady grove. Food, fruit, ice cream, and water sprinklers awaited. By the time we set off in the afternoon, we were about ten thousand marchers, walking with great enthusiasm and elated spirits. When we arrived that evening in Kibbutz Nachshon, we were around twenty thousand marchers, with more people flocking in. There were food and drinks on tables in what had turned into a full festival, with speakers on stage, live music performances, flowing beer, and an endless outpouring of kindness and care toward one another—all of which brought much hope, which is vital for any form of political activism (Leshem et al. 2024).

The following day, the number of marchers doubled. In the evening, more people joined the camp organized in a forest, ten kilometers outside Jerusalem, where they drank wine and sang Shabbat songs, among other tunes. For the final stretch, on Saturday, July 22, we were more than seventy thousand marchers walking uphill on the main road toward Jerusalem, where many supporters were waiting to welcome us. The march ended after we circled the parliament, and eventually thousands stayed to camp near the Knesset for two days, waiting for the vote to take place. The full outcomes of the vote are beyond the scope of this article, but I will note that the coalition managed to pass the law on the unreasonableness clause. Some appeals against this law were brought before the Supreme Court, which, in a historic ruling months later, struck it down. According to many commentators, this bold ruling was possible because the judges could morally rely on the massive public mobilization against the law in July 2023.

The July march was marked by many marchers carrying the Israeli flag. The reclaiming of the national flag had occurred some months earlier as a strategic move to combat questions of loyalty. As a result, during the march, it was impossible to see through the sea of people and their flags to estimate how many marchers were ahead of or behind you. At any given moment, you could only see a few dozen marchers. This meant that, while marching with great attention to the profusion of bodily movement and enthusiasm, the "big picture" was always out of sight. This is reminiscent of de Certeau's classic discussion in the often-quoted chapter on "Walking in the City" (de Certeau 1984), where he distinguishes between bodily language and scopic regimes. As a Jesuit, and inspired by the figure of the *Wandersmann*, de Certeau favors partial bodily language over the temptation of a totalizing view of the world from above. As Charles Lock (1999) notes, the latter is reminiscent of Satan's temptation of Christ to see the entire world from an elevated vantage point (Luke 4, 5–8; Matthew 4, 8–10).

De Certeau wrote his work in 1980, but since then, technology has made the choice between these two modes of knowledge less relevant: one can now know with the body and from above at the same time. Since the inception of the protests against the government, an entire crew of photographers has become integral to the movement. They photograph demonstrations with the dual purpose of pushing images to the media and documenting violent arrests by the police. The photos can later be used to contest those arrests, demonstrating how (almost all) had no legal grounding, while also forming the basis for claims against police violence. The photographic arm of the movement shares an online drive, to which I was granted access as part of this research. In addition, a drone squadron has been used for similar purposes—that is, to show how massive the demonstrations are and to use the vertical dimension as an aesthetic tool to encourage more people to take to the streets. Often one would demonstrate in Tel Aviv with four drones hovering overhead. Beyond photography, and drone photography in particular, since the protests began in 2020, there has been a proliferation of WhatsApp groups. As a result, activists in this movement are bombarded with dozens—sometimes hundreds—of messages a day from different groups, many of which are active during protests.

This technological Zeitgeist—the rapid spread of photography, particularly aerial photography, within WhatsApp groups of demonstrators—brings with it a set of advantages. However, it also evokes a stage in global surveillance, where various forms of monitoring are not only carried out by state apparatuses but have also become part of our surveillance culture (Lyon 2017). We take it for granted that we are potentially being photographed, and that our activities could be transmitted through social media to various stakeholders, both friends and foes. We may also participate in this process. This awareness had a profound effect on the July 2023 march. At any given moment, one could see only a few dozen marchers, the rest hidden behind the crowds, their fists raised to the sky and flags waving. Yet, at the same time, one could look at a smartphone and see livestreams or photographs from just minutes before, showing long shots or vertical views of the entire group of marchers. As you marched, people around you would show you an image on their phone, saying, "This is how we look!" In this sense, even the gait of the march was carried out with both the body and the knowledge that someone was watching you from a distant, elevated point. The intensity of bodily knowledge and the immense sensory effects were in constant interplay with the view from above. In contrast to the walker that de Certeau imagined—whose footsteps are like speech-acts, fragments of a text the walker cannot read—here one wrote the text and read it simultaneously, consciously writing to be read.

Furthermore, the July 2023 march was consciously associated with the myth of Exodus. Escorting vehicles involved in the logistics of the march played "A Man Gets Up in the Morning," a 1971 hit by the Israeli singer Shlomo Artzi, known for its first verse, which translates as "Suddenly a man gets up in the morning feeling that he is a people, and he begins to walk / And to whoever he meets on his way he says hello (shalom)." Originally a poem written by Amir Gilboa in 1951, the song has acquired anthem status, drawing a direct connection between the personal, mundane act of getting up and walking and being part of a nation. Gilboa himself explained in an interview that this walk stood in opposition to that of the Wandering Jew (Hasan-Rokem and Dundes 1986). In his view, this was a walk brimming with agency after two thousand years of dormancy. The rhythmic quality of Artzi's song added an important dimension to the march, and with the words playing from mobile loudspeakers, it reinforced the idea that this personal march was being harnessed to (re)build a just nation.

One of the key figures in initiating the march was Moshe Radman, who emerged as a leading voice within the 2023 demonstrations and on social media. He soon became the face of the march: with a white shirt tied around his head and a wooden staff in his hand, people began referring to him as "Moshe Rabenu" (Moses, our leader). Drone footage captured the marchers in impressive long shots, reminiscent of Cecile B. Demille's *The Ten Commandments*, with its grandiose vision of the Israelites wandering in the desert. One video of the march, taken by a traffic camera mounted on a pole and uploaded to YouTube, was explicitly titled "The Exodus from Egypt on the Road to Jerusalem." The march took place in extreme heat, emphasizing the hardship that had to be endured—again evoking the Exodus narrative, where such struggles played a vital role in the formation of an independent people. The marchers chanted the entire walk, holding their flags high, with both the fast pace and verticality of the body contributing to their demonstration of power.

The marchers kept multiplying from one day to the next, just like in the biblical story of the Israelites in the desert, though no Manna was falling from the sky. An entire parallel operation to the march itself was carried out in an attempt to feed the tens of thousands of marchers along the way. The logistics were handled by volunteers who kept arriving, and for the marchers, this was perhaps the most moving aspect: food and drink were being brought to them not just for survival—the act was an expression of love and camaraderie. When the march was announced, very few preparations had been made, and the incredible "catering" along the way was shaped entirely by volunteers, under the supervision of Ayelet Katzir, who stepped up for the occasion rather spontaneously. Her own story is particularly telling (Katzir, interview with the author, May 7, 2024). A former kindergarten teacher and a senior kindergarten supervisor in the Ministry of Education, she had retired early after a serious sports injury. As a result, she couldn't march herself, but she wanted to be involved all the same. On the first morning of the march, she packed a few things and a picnic table, thinking she could help support the march without walking. When she arrived, however, and saw the number of marchers surpassing all expectations, she was approached by some of the organizers and asked if she could take charge of the logistics. With no prior knowledge, she drafted a list of needs and recruited her family to assist. Katzir orchestrated the entire operation: breakfasts, lunches, dinners, and fruit, ice cream, and drinks in between. No detail was overlooked: she made sure there were tables for marchers with gluten-free diets, vegans, and those keeping kosher. "Nobody could be left out," she told me. This attitude stood in stark contrast to the disappointment felt toward a government that was expected to care for its people but had failed to do so.

In addition to Katzir's unique organizing skills (which she only discovered she had during the march), many volunteers arrived with provisions. A buffet for tens of thousands of marchers was ready at every stop. At every turn, I was offered fruit or ice cream—once, twice, sometimes four times. This was very much an exercise in generosity and abundance that mirrored the exercise in strength of the actual march. This bottom-up celebration of caring for one another was often seen as the antithesis of the selfish, power-driven, and corrupt government.

Despite the political outcome, the marchers left with a euphoric feeling of belonging. It was no coincidence that, in the months that followed, similar smaller marches took place, like one in the Galilee that drew fewer people but the same abundance of altruism—marching in forty-three degrees Celsius under the scorching sun. This march, which happened in September, was meant to be the second in a series. "We shall march everywhere and meet people. We shall meet people on foot and rebuild this country completely," one marcher declared in a speech before us at the end of the march.

November 2023: Marching as a Funeral Procession

Things took a completely different turn a few weeks later with the October 7 attack. It started very early in the morning for those in the kibbutzim near Gaza and for those raving at the Nova music festival. For others, it began when they woke up—either because rockets were fired at their area or because they were directly contacted by those who had been attacked. For me, it started two hours after the attack began, when rockets aimed at Jerusalem triggered the sirens that woke my family immediately. An hour later, all the members of our university folklore program were WhatsApping each other: three students lived not far from Gaza, and Kamelia Hoter-Ishay lived in Be'eri, which, in retrospect, was the epicenter of the attack (Kamelia asked to appear in this article with her full name).

The rest of that Saturday, we received horrifying messages from Kamelia's security room, as she grappled with even worse messages from her two daughters, who are full members of the kibbutz. A few days after the attack, we attended the funeral of her son-in-law and waited for news about her grandchildren, who were missing. A week after the attack, the body of her grandson was found in his home, which had been burned to the ground. When November began, Kamelia and her family received the devastating news that her thirteen-year-old granddaughter, G., who was missing, was actually kidnapped (she was the last to be reported as such). Kamelia and her daughter became very active in the tent for the families of the hostages in Tel Aviv, campaigning for negotiations to bring back their loved ones. As part of that effort, on November 14, 2023, these families launched a five-day march under the slogan, "Walking to bring them back!"

When I arrived at the encampment of the families in Tel Aviv, from where the march would set off to Jerusalem, Kamelia asked me to march with a large sign featuring the image of G., which I carried for the next few days. The comparison to the march that had taken place four months earlier began with my very first steps. While the first march held two perspectives at once—a bodily experience and a vertical scopic one—here every gesture was overshadowed by a strong sense of déjà vu or "déjà marché"—of walking two marches at once. I had the chance to discuss this with several marchers who took part in both marches, and they all shared similar feelings. Comparisons point to similarities but also to differences, and here those differences were so striking that the weight of the July march made the heavy, sorrowful atmosphere of the November march even more palpable.

The families led the way, each holding signs with pictures of their loved ones. Few marchers carried Israeli flags, and no songs were sung. Every now and then, we would call in unison, "Bring them home now!" Whereas the July march was the result of months of demonstrating, after the October 7 attack, activism took on a different form, with aid centers and volunteering becoming central. As mentioned, activism is fueled by hope, and the feelings of despair and the atmosphere of mourning kept many indoors. The fact that many were too depressed to move may explain why the number of marchers, which grew daily during the five-day march, only reached around twenty thousand on the final stretch uphill to Jerusalem.

Although the point of departure and the destination were the same, the route was not identical. Instead of taking the shortest possible path to Jerusalem, the November march meandered through villages and municipalities, where the marchers were greeted—sometimes ceremonially—by local activists, politicians, and residents expressing solidarity with the families and participants. At each stopover, food and drink were provided by the local hosts. Members of the local organizing team would typically speak through a sound system to the marchers, welcoming the families and marchers. Then representatives of the families would deliver their own short speeches and pleas for the release of their loved ones. In general, the families not only led the way but were also given separate zones to stay in. In that sense, both the hosting municipalities and those joining the march (like myself) saw it as their moral duty to be there for the families, expressing solidarity with their deep anguish and pain by opening their hearts and physically walking together.

Except for a rainstorm on the first day, the weather was cloudy and comfortable for walking. The explicit hardship of marching under a scorching July sun was replaced by a different form of bodily sacrifice: the march was deliberately executed at a very slow pace. Because the families leading the march set the rhythm, they walked slowly, often stopping to pose for photographers eager to capture as many images of the poster signs of the kidnappees as possible. This imposed rhythm, combined with the grief that pervaded the march, made the entire procession feel more like a funeral march. Each step began with a slow lift of the ankle, almost dragging the toes in very short steps. Marchers kept their heads down, walking hunched over, and after a few hours, their lower backs ached with terrible pain. Holding the poster signs against the wind added another physical challenge, as, unlike flags made of a flexible fabric, the signs were often mounted on wooden plaques.

The faces of those kidnapped have become familiar to many Israelis, who see them at bus stops, squares, and public institutions. In other words, the signs have operated as icons, with the faces of those missing becoming familiar public imagery (Belting 1994). As I held the poster sign of G., it became almost a personal mission to keep it intact during the rainstorm and in the face of the wind. I held the poster as high as I could, so it would be visible, imagining that this would keep her in the public conscience and prevent her from being forgotten. In this way, the act of walking with the signs was reminiscent of Orthodox Easter processions, where icons and crosses are held high. Because those holding the poster signs were mostly relatives or close friends of the kidnappees, I was often treated as though I knew G. or was related to her. This resulted in some awkward situations where people urged me to go to the front of the march, where the families led the way. I had to explain my status, though sometimes there wasn't time—like the moment when a man in his sixties saw me holding the sign, approached me, and hugged me, nearly crying on my shoulder. When we arrived in front of the parliament for the final gathering, a woman saw G.'s face on the poster I held firmly and deliberately as high as possible. She approached the sign, calling out G.'s name through tears, almost hugging it as though G. were her own daughter, sending kisses to her in a devotional way—again reminiscent of practices of icon veneration (Pentcheva 2006).

Shortly after this march, the Israeli government and Hamas leadership negotiated an agreement in which some kidnappees were released daily. Late at night on November 29, G. was seen for the first time after a month and a half, stepping out of a jeep on the Egyptian side of the border with a gigantic smile that the media fell in love with. The talks between the Israeli government and Hamas ended sourly. At the time of writing, these talks were renewed and the fate of many of the hostages and those kidnapped is still unclear, as is the situation in Gaza, and the state of war, with its endless casualties—Palestinians in Gaza in particular—is ultimately discussed behind closed doors by international leaders.

Comparing Practices and Walking Genealogies

The idea that one can march to achieve certain political ends has several histories (for example, Dinkel et al. 2024). It is a transnational history and, in fact, the title "political marches" encompasses different techniques of the body (Mauss 1973), distinct symbolic languages (Merriam 1975), and various walking models that inform these marches. Marches have been used by modern armies to facilitate order as part of the control over docile bodies (Foucault 1978), ultimately molding them to move in unison ("left-right-left"). This demonstrated the potential power of a trained army. Yet this control of movement also created distinctions; for example, as Mauss observed, the gait of British soldiers was incompatible with that of French soldiers (Mauss 1973, 72). Religious processions, where walking bodies are ordered and coordinated in ritual, also form part of this history (Amato 2004, 44). The discipline exhibited in both the church and, of course, the army reflects well-established hierarchies and power structures.

The marches considered here, on the other hand, are based on the association of marchers whose social activity harnesses bodily forces to create a collective power against existing power structures. Unlike collective listening or spectatorship, marching is an immensely immersive activity that is physically demanding and politically binding in ways that foster strong collective identification. As noted, from my first step in November, my mind was constantly comparing the two marches, and the persistent feeling of déjà vu gradually gave way to a deep acknowledgment that the two marches were entirely different experiences with very different bodily languages and movement repertoires. I slowly realized that the idea of a political march, which I had taken for granted as relatively stable, requires much more careful examination.

In his discussion of workers' marchers in the German context, Bernd Jürgen Warneken (1986) explains that such marches drew on the experience of military Prussian marches, but he distinguishes between these marches and *Massentritt*, which can be translated as "mass-tread" or "masses-step." Walking and modernity are often discussed in relation to Walter Benjamin's interpretation of Charles Baudelaire's flâneur—an emblematic figure of modernity who stood in opposition to the movement of the masses (Benjamin 1999; Gluck 2003). The flâneur's potent walk often countered the current of the masses; his creative and unpredictable manners challenged modern conformity. (His female counterpart, the flâneuse, was much less powerful—see Wolff 1985; Iskin 2003.) However, the power of the masses walking together became a more common practice in modernity, emerging as part of what it meant to "feel political" (Frevert 2022).

The key difference between how political marching evolved, compared to military marches, lies in the way bodies are set in motion with force that is not violent. This is tied to Mahatma Ghandi's marches and his conception of *Satyagraha* and nonviolent resistance, which had a profound impact on South Africa, India, and the civil rights movement in the United States. These marches—with their overt nonviolence—gestured toward power in symbolic ways (Merriam 1975; Siganporia 2022; Maul 2024). Ghandi's bodily performance is a classic example of the politics of exemplarity recently discussed by Dorothy Noyes (2024), but so is every participant who joins the march. An excellent indication of the complexity of marching cultures is the different words used to describe these practices, often with semantic fields that are not consistent, especially in translation—terms like "slut-walks" (Möhring 2024), death marches, May 1 parades, March of the Living, gay parades, army parades, carnival parades, and Easter processions. The actual differences in the physical acts of walking correspond to the many words used to describe these collective walking practices.

I joined these two marches with my own inclinations and expectations. This became even clearer to me when I joined another march in support of the hostage families in February 2024. By coincidence, it took place during the same week I was teaching a class on marching in my walking seminar, where we were reading the works of Ute Frevert, Bernd Jürgen Warneken, and Harmony Siganporia, which I discussed earlier. As I shifted my attention between bodily memory and scholarly reflection, I could see how the readings affected my marching and how marching, in turn, affected my textual interpretation. In July 2024, I joined another march with the families—tens of thousands once again under the summer sun. By then, my senses and my body were trained to differentiate between the different performances of the march. On that day, Israel launched another attack in Gaza in an attempt to assassinate one of the leaders of Hamas, killing many Palestinian civilians. Later that night, Donald Trump narrowly escaped an assassination attempt. As a result, this very noisy and energetic march went almost unnoticed by the media; news editors probably had trouble differentiating between these marches from Tel Aviv to Jerusalem, which, sadly, from their vantage point, had become routine, much like the grief of wars. The power of this particular march, and the knowledge of marching, remained stored in the bodies of those who took part.

And That Has Made All the Difference

In the first march, the bond between marchers was based on shared myths, common goals, and dreams for the future. The second march, in contrast, brought walkers together around human tragedy and the nearly impossible task of continuing to live in the present. Hierarchies shifted accordingly—in July 2023, the leaders of a protest movement were paving the way for the future, while in November, those closest to the tragedy, whose own pain was the most extreme, leaned on a supporting network in their march. These evident differences of circumstance were marked not only in the different steps and bodily adaptations of the two marches but also in the constant reflection and comparison between the marches made by those whose bodies were shaped collectively.

Marches can mobilize people for specific goals, and in both cases, these goals were partially achieved. The judiciary overhaul is still underway, but there is now a balance of power between the government and a public that is much more alert. Marches have become crucial in building a long-term civic movement against the government. This movement is based on continuous learning and critical reflection—reflection that takes place in words and discourse, but much of which is carried out in the memory of the feet.

References

Amato, Joseph. 2004. *On Foot: A History of Walking*. NYU Press.

Belting, Hans. 1994. *Likeness and Presence: A History of the Image before the Era of Art*. University of Chicago Press.

Benjamin, Walter. 1999. *The Arcades Project*. Translated by Howard Eiland and Kevin McLaughlin. Harvard University Press.

de Certeau, Michel. 1984. *The Practice of Everyday Life*. Translated by Steven F. Rendall. University of California Press.

Dinkel, Jürgen, Kai Nowak, and Miriam Pfordte, eds. 2024. *Märsche der Moderne: Varianten eines globalen Phänomens*. Campus Verlag.

Foucault, Michel. 1978. *Discipline and Punish: The Birth of the Prison*. Translated by Alan Sheridan. Pantheon.

Frevert, Ute. 2022. "Feeling Political in Demonstrations: Street Politics in Germany, 1832–2018." In *Feeling Political: Emotions and Institutions since 1789*, edited by Ute Frevert, Kerstin Maria Pahl, Francesco Buscemi, Philipp Nielsen, Agnes Arndt, Michael Amico, Karsten Lichau, et al. Springer International Publishing.

Gluck, Mary. 2003. "The Flâneur and the Aesthetic: Appropriation of Urban Culture in Mid-19th-Century Paris." *Theory, Culture & Society* 20 (5): 53–80. https://doi.org/10.1177/02632764030205003.

Hasan-Rokem, Galit, and Alan Dundes, eds. 1986. *The Wandering Jew: Essays in the Interpretation of a Christian Legend*. Indiana University Press.

Iskin, Ruth. 2003. "The Pan-European Flâneuse in Fin-de-siècle Posters: Advertising Modern Women in the City." *Nineteenth-Century Contexts* 25 (4): 333–56. https://doi.org/10.1080/0890549032000167844.

Kremnitzer, Mordechai. 2023. "Releasing the Government from Acting Reasonably; or, the Government Says Goodbye to Reasonableness." *Israel Law Review* 56: 343–54. https://doi.org/10.1017/S0021223723000146.

Leshem, Oded Adomi, Shanny Talmor, and Eran Halperin. 2024. "Hope and Activism." In *Hope Amidst Conflict: Philosophical and Psychological Explorations*. Oxford University Press.

Lock, Charles. 1999. "Michel de Certeau: Walking the via Negativa." *Paragraph* 22 (2): 184–98.

Lyon, David. 2017. "Surveillance Culture: Engagement, Exposure, and Ethics in Digital Modernity." *International Journal of Communication* 11: 824–42.

Maul, Daniel R. 2024. "Peace Is the Way. Gewaltfreiheit als Prinzip und Methode in Gandhis Salzmarsch 1930 und dem Marsch auf Washington 1963." In Dinkel, Nowak, and Pfordte (2024).

Mauss, Marcel. 1973. "Techniques of the Body." *Economy and Society* 2 (1): 70–88. https://doi.org/10.1080/03085147300000003.

Medina, Barak. 2023. "Legal Challenges of Mass Demonstrations: The Case of Israel." *SSRN*. https://papers.ssrn.com/sol3/papers.cfm?abstract_id=4582070.

Medina, Barak, and Ofra Bloch. 2023. "The Two Revolutions of Israel's National Identity." *Israel Law Review* 56: 305–19. https://doi.org/doi:10.1017/S0021223723000225.

Merriam, Allen H. 1975. "Symbolic Action in India: Gandhi's Nonverbal Persuasion." *Quarterly Journal of Speech* 61 (3): 290–306. https://doi.org/10.1080/00335637509383294.

Möhring, Maren. 2024. "'Come walk or roll or strut or holler or stomp with us'. Der Slutwalk als globaler feministischer Protestmarsch?" In Dinkel, Nowak, and Pfordte (2024).

Noyes, Dorothy. 2024. "Folklore, Exemplarity and Politics." *Voices of Excellence* podcast episode posted June 26, 2024, by Ohio State University College of Arts and Sciences. YouTube, 29:54. https://www.youtube.com/watch?v=jliYAjr3lC0.

Pentcheva, Bissera. 2006. "The Performative Icon." *Art Bulletin* 88: 631–55.

Siganporia, Harmony. 2022. *Walking from Dandi: In Search of Vikas*. Oxford University Press.

Warneken, Bernd Jürgen. 1986. "'Massentritt': Zur Körpersprache von Demonstranten im Kaiserreich." In *Transformationen der Arbeiterkultur*, edited by Peter Assion. Jonas Verlag.

Wolff, Janet. 1985. "The Invisible Flâneuse. Women and the Literature of Modernity." *Theory, Culture & Society* 2 (3): 37–46. https://doi.org/10.1177/0263276485002003005.

12

Walking Outside the Box

A Visual and Sensory Landscape Journey to Capture Perceptual Values

Dominica Williamson, Kayla Parker, John Martin, and Joane Serrano

This chapter proposes a conceptual shift in our perception of "landscape" as a cultural term to move beyond the colonizing paradigm of the West (Smith 1999). It advocates for a more holistic approach, rooted in experiential knowledge, to augment the customary assessment of landscapes, which typically focuses on physical, socioeconomic, and aesthetic features. The authors propose the "mini-landscape" method (Williamson et al. 2016; Martin et al. 2021) as an example of how perceptual landscape data can be collected and evaluate this innovative approach within a new materialist framework. The method aims to understand how people perceive and value landscapes, arguing that physical presence within an environment is essential, as bodily presence significantly influences perception.

The mini-landscape method centers on workshop participants sharing their experiences of walking as a group through a landscape to a destination. Upon arrival, they gather and assemble materials found at the site—such as feathers, rubbish, soil, an empty snail shell, and fallen leaves—to create imaginary, miniature landscapes within a bounded area, like a vivarium or a designated space. This interdisciplinary method, developed through a series of workshops over an eight-year period in rural and urban landscapes in the United Kingdom, several North European countries, Southern Italy, and the West Indian Ocean, has involved a range of stakeholders, including adults and schoolchildren from various backgrounds and generations. The mini-landscape method offers a phenomenologically informed approach to revolutionizing how people view and understand their environment.

The discussion is grounded in posthuman understandings of the more-than-human world, drawing on the works of David Abram, Edward Relph, Erin

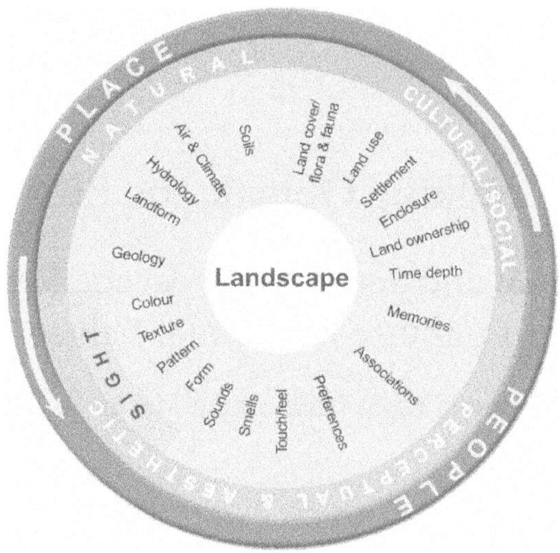

Figure 12.1 Landscape Character Wheel. Source: Permission to print approved by Natural England (Copyright).

Manning, and others, and is informed by Indigenous knowledges, such as those of Linda Tuhiwai Smith. With contributions from workshop participants, it outlines the collective, evolutionary nature of the mini-landscape method, its adaptability, and its proven efficacy in capturing perceptual data, taking the reader step by step through the workshop activities and processes in the field. The Landscape Character Assessment (LCA) method used by Natural England (Tudor 2014), the UK government's adviser for the natural environment, and its LCA wheel (figure 12.1) provide a useful comparative model for advancing the mini-landscape method, which foregrounds experiential knowledge through multiple senses, including touch, smell, taste, and proprioception. The chapter argues for a blurring of the rigid categorizations of the LCA wheel, proposing the inclusion of additional features to capture perception at a deeper level. It highlights the imperative for us to move beyond the rigid confines of the Eurocentric model of landscape assessment—"walking outside the box," listening to the voices of the Global South and Indigenous peoples, and embracing the "messiness" of perceptual data.

Holistic and Sensory Approaches to Landscape

The character of a given landscape is influenced by both nonhuman and human factors, reflecting their multifunctionality and supporting the concept of a sustainable and cultural paradigm for the twenty-first century (Hermann et al. 2011). A landscape's character can also evolve over time, in response to the continual movement of human societal preferences and natural processes (Council of Europe 2000). In developing policies to protect and manage landscapes, government organizations may use various characterizations and assessment processes.

The LCA methodology emphasizes the need for the public to play an active role in the characterization process, as perceptual and aesthetic elements are linked to stakeholder engagement. Within this process, the guidelines suggest engaging with two types of communities: those of "place" and those of "interest." The European Landscape Convention (ELC) mirrors the Natural England approach, defining landscape as "an area, as perceived by people, whose character is the result of the action and interaction of natural and/or human factors" (Council of Europe 2000).

A holistic approach is therefore preferred in the characterization process, encapsulating the diversity of qualities contributing to the "sense of place" (Landscape Institute 2016). Place plays a central role in human experience. Relph (1976) points out that human beings have a deep need "for associations with significant places"—it is important to us to connect with place and to "transcend placelessness" in order to develop environments for people that reflect and enhance the variety of human experience (1976).

Although the LCA refers to people and the importance of capturing values, in practice it often lacks bottom-up approaches that would ensure stakeholder contributions. This may be due to the challenges of gathering "perceptual data" compared to collecting data from other aspects of the environment indicated in the landscape wheel. While soil maps and land ownership records offer reliable sources of data that can be perceived through sight, perceptual data involves a multitude of senses. These include touch, smell, and taste, as well as other senses that are not fully understood and are difficult to quantify, such as proprioception.

Alva Noë argues against the traditional view of the senses as separate and independent faculties, as demonstrated in his work with J. Kevin O'Regan in 2001. In contrast, Jerry Fodor's (1983) work often underscores the modularity of cognitive processes. However, Thomas Metzinger (2009) acknowledges the importance of the body and environment in shaping conscious experience, though he prioritizes a neuroscientific understanding of consciousness—a perspective that diverges from Noë's emphasis on embodiment and enaction. Considering these differences, Ryan Putt (2016) highlights the subjectivity and temporal dependency of perceptual data, noting that not only is each person "feeling" the landscape differently but each season and each day is different, and even the weather can change from hour to hour. The LCA, in addition to being compartmentalized and not effectively capturing perceptual data, is also limited to anthropocentric definitions and processes.

A Call to Return to Sensoria

The notion of reexamining the centrality of human positions resonates with Indigenous knowledge systems in the Philippines and other Asian contexts. For example, many Indigenous communities in the Philippines believe in *diwata* (nature spirits) and *anito* (ancestral spirits), whom they regard as guardians of specific natural elements. These beliefs underscore the deep respect Indigenous peoples have for the interconnectedness of humans, animals, plants, and spirits, reflecting a worldview that places nature and spirituality at the heart of their communities and fosters harmony between humans and the natural world.

The term "more-than-human world" was introduced by David Abram (1997). He articulates a conceptual repositioning of human beings' relationships with nature that ruptures the binary model of "us and them"—or "us and it"—in which "humans" are separate and dominant in relation to all "other" animate and inanimate entities that share our planet. Abram's reappraisal of the philosophic centrality of the human position draws on Maurice Merleau-Ponty's (1962) focus on the "sentient and sensuous" body as the "experiencing 'self'" and on Indigenous knowledge. He thus challenges the notion of humans as superior to, and "masters" of, the natural world and its resources. This radical conceptual realignment creates spaces in which other imaginaries can take root and flourish. For example, Indigenous cultures and customs are integral to our approach in developing the proposed research method, as they contribute non-Western, decolonizing, and ecological approaches to geography, land management, and biopolitics. These voices are of critical relevance during this burgeoning period of anthropogenic climate emergency, as exemplified by the foundational research of influential Māori scholar Linda Tuhiwai Smith in *Decolonizing Methodologies: Research and Indigenous Peoples* (1999).

Abram (1997) emphasizes the importance of relearning our reciprocal relationships with nature, encompassing not just flora and fauna but also inanimate elements like rivers and mountains. New materialist thinking and

Indigenous ways of knowing extend this understanding, providing a foundation for environmental ethics. These perspectives inform the mini-landscape method. By incorporating raw, natural materials, the method fosters a deep sensory connection and cocreative relationship with the environment. Participants engage with landscapes on a tactile level, allowing them to explore not only the physical space but also their sensory and emotional connections to it. This approach reflects the Indigenous understanding of a reciprocal relationship with the natural world, viewing landscapes as dynamic, living assemblages rather than static objects.

It is important to note that while the decolonization of Indigenous knowledge systems, as discussed by Smith (1999), often serves as a form of resistance against Western influence, not all Indigenous practices can be framed solely as acts of resistance. For instance, the Ifugao Rice Terraces are not a response to colonial systems but rather a product of Indigenous ingenuity and sustainable cultural practices. These systems were developed over centuries as part of the Indigenous peoples' intrinsic relationship with nature.

Building on new materialist approaches, such as more-than-human and sensory perspectives, alongside Indigenous knowledge, we argue that it is essential to move beyond the "communities of place" and "communities of interest" identified in the LCA approach and to introduce "communities of nature" (as suggested in figure 12.2). This shift would contribute to decompartmentalizing the wheel: rather than adding more segments, we need to blur the demarcation lines in order to address the complex, messy data needed to understand landscape character.

Methodological Summary

In addition to advocating the need to address "communities of nature," the method promotes an approach that emphasizes the sharing and collection of perceptual data. The proposed mini-landscape method is an immersive, participatory activity that invites groups to explore their landscape and cocreate a miniature representation, reflecting their collective perception of the space. The method highlights the importance of physical presence within an environment, as bodily presence significantly influences perception (Tilley 1994; Pink 2015). Embodiment theory suggests that cognition is closely tied to the body, meaning that sensory experiences, movements, and interactions with the environment

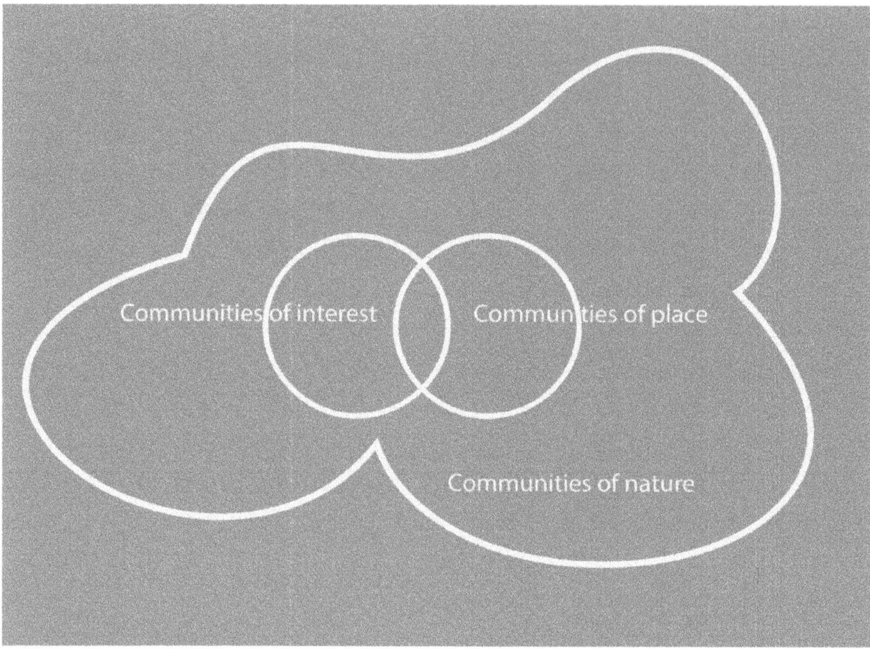

Figure 12.2 Communities of nature. Source: Illustration by Dominica Williamson.

are integral in shaping individuals' mental processes, including perception, cognition, and emotional responses (Merleau-Ponty 1962; Lakoff and Johnson 1999). The mini-landscape method thus advocates for embodied experience within the landscape as a primary means for collecting perceptual data, aligning with other methodologies that emphasize an embodied engagement with place (Ingold 2000; Abram 1997). The method engages with three embodiments over three stages of activity, resulting in outputs at each stage, as detailed in table 12.1.

As previously mentioned, these workshops have been conducted over the past eight years with a range of stakeholders, including adults, youth, and schoolchildren from various backgrounds and generations. The methodology, codeveloped in collaboration with different global communities through various research projects, incorporates insights from the field of visual and creative methods, as advocated by scholars such as Aline Gubrium et al. (2016). Characterizing this method is challenging, as referring to it as a "visual method" could imply an overemphasis on vision, while calling it a "creative method" doesn't necessarily foreground the perceptual. As a result, we present this approach as a "perceptual method," which aligns with new materialist

Table 12.1 Mini-Landscapes: A Perceptual Method for Cocollecting Landscape Values. The method aims to understand how people perceive a landscape, establishing that physical presence within an environment is essential to this process, as bodily presence significantly influences perception.

Embodiment	Steps	Output
Stage 1: Moving through the landscape This stage has been developed and refined by engaging in walking through a landscape.	1 Workshop arrangements: - To prepare, a walk is planned in collaboration with a landowner/manager or through discussions with participants familiar with the landscape. - Walk duration: fifteen to forty-five minutes, tailored to accommodate varying mobility needs. 2 Workshop day: - The group of approximately ten to fifteen participants is greeted upon arrival. - The facilitator (or facilitators) introduces everyone and gives a brief talk about the landscape and the destination. 3 The walk: - Participants explore landscape types along the route as a group. - Participants talk about the landscape features and share their thoughts, memories, and information about the site. 4 Destination area: - Participants arrive at the prearranged site where the tools for creating the mini-landscapes are provided, such as vivaria, trowels, bags, and buckets.	As participants move through the landscape, their bodily experiences, movements, and sensory perceptions are influenced not only by the environment but also by the presence of and interactions with others within the group.

(Continued)

Table 12.1 (Continued)

Embodiment	Steps	Output
Stage 2: Creating mini-landscapes Each person works within a subgroup to create a mini-landscape.	5 Division into subgroups: - Four to five subgroups are formed, each consisting of two to four participants. - Each subgroup is tasked with constructing a mini-landscape in a vivarium or designated area. 6 Discussion: - Within the subgroups, participants discuss what form the mini-landscape should take. Options might include an abstract, narrative, simulated, or literal representation of the workshop landscape itself. 7 Construction: - Subgroups have one hour to gather and arrange materials to create their mini-landscape within the vivarium. - They may choose to place the vivarium carefully in a different part of the workshop site. 8 Facilitator support: - Facilitators visit each subgroup to provide help or advice as needed.	Mini-landscapes become “more-than-individual” expressions, reflecting a collective understanding of the landscape as coperceived by the subgroup.
Stage 3: Inscribing, laying down, and uttering the cocreation Participants write one or two words on glass microscope slides, record their mini-landscape using their phones, and then share their experiences with the group.	9 Questions: - Facilitators ask each subgroup to provide answers to eight open questions about the landscape, using eight glass microscope slides and a glass pen (one to two words per slide). - The questions relate to the eight senses: visual, auditory, olfactory, gustatory, tactile, proprioception, vestibular processing, and interoception.	The open questions encourage deeper engagement with the process. Participants are encouraged to document their own work and experiences. Facilitators and participants exchange ideas, blurring the lines between researcher and participant so data is cocreated and shared.

10 Sharing slides:
- Each person places their slides on the ground, sharing them with their subgroup.
- Each subgroup selects slides to display in or around their vivarium.

11 Examining the vivaria:
- The vivaria are placed close together, if possible, so the entire group can view the mini-landscapes that have been created.
- Participants examine each vivarium closely.

12 Summary:
- Each subgroup presents a summary of their mini-landscape to the entire workshop group.
- A record of each summary is made through photographs, sound recordings, and videos.

13 Keeping a record:
- Participants are encouraged to take photographs of their mini-landscapes with their phones before dismantling them or "growing on."
- Facilitators guide participants in using various photographic techniques. For example, for close-up shots, participants can lie on the ground or use a phone rest for stability.

Figure 12.3 A cocreated mini-landscape. Source: Photograph by Andy Hughes.

research assemblage (Fox and Alldred 2014). In this assemblage, every aspect of the workshop, both physical and philosophical, must be considered alongside what already exists in the workshop space. Every element, including the "raw material" of the landscape, contributes to shaping and influencing the outcome of the research.

New materialism offers a nonanthropocentric framework that allows us to disrupt the traditional duality of categorizing people and nature as separate, with nature reduced to "capital"—a resource to be exploited. It allows us to understand "landscape" as an interrelational nexus of different places, where the meaning of a landscape is rooted in people's engagement with those interconnected places—in what they do there, how they feel about them, and how their experiences of the landscape, its associations, and its histories intertwine with the events that have occurred there.

For Abram, the living, corporeal, material body is an entity whose porous boundaries are "open and indeterminate; more like membranes than barriers, they define a surface of metamorphosis and exchange" (1997, 38). This body-centered paradigm situates our perception and phenomenological experience within the more-than-human world that surrounds us rather than positioning us as separate from it, and acknowledges our "selves" as coparticipants, intra-acting with all other "things."

The emerging field of eco-phenomenology applies new materialist thinking to address and critique our current relationship with nature and the environment. It advocates for a process-based approach to fieldwork that supports the evolution of the perceptual data-gathering methods embedded within the mini-landscape workshops. The perceptions contributed by workshop participants form a rich and holistic body of data, embracing a diverse array of responses to the landscape experienced during the workshop. This perceptual data provides insights into what Abram, citing Merleau-Ponty, calls "a silent conversation . . . with things, a continuous dialogue" that unfolds beyond our verbal awareness and "an improvised duet between [our] animal body and the fluid, breathing landscape it inhabits" (Abram 1997, 53).

In this sense, the workshops may be understood as "making place"—facilitating participants' deep engagement with the specificities of the places that constitute the landscape, allowing them to create meaning. The sites chosen for the workshops vary from woodlands and cultivated forests to urban parks—each revealing its phenomena through the researchers' and the project participants' bodily experiences of these places. The mini-landscape method allows the participants' senses to connect with what lies beneath the visible landscape, capturing experiences and meanings that often go unnoticed in everyday life (See figure 12.3 for an example of a cocreated mini-landscape.).

Through field practice, the method facilitates the production of perceptual data that can be analyzed. Importantly, this approach acknowledges the significance of embodied experiences and intra-actions with the material world, in line with new materialism, which highlights the agency of more-than-human entities in shaping human experiences. The method is sensory and inclusive, foregrounding the materiality of the encounter—encompassing the materiality of the human body, its corporeality, and its psychological and affective dimensions, as well as the materiality of the places in the landscape through which we move, embracing everything that is "not us"—that is, "other than human": the more-than-human.

This conceptual shift and its exploration are now followed by an expansion of the mini-landscape method through the lens of three forms of embodiments (as highlighted in the first column of table 12.1).

Embodiment 1: Moving through the Landscape

"The peripatetic being can no longer be considered as a solitary walking figure enclosed in private thoughts. A whole world of historical thing-relations participates

Figure 12.4 Workshop participants moving through a landscape. Source: Photograph by Jason Parsons.

in the making of one's walking rhythms. The person on the move is intimately making contact with the rich fabric of social formations which are sedimented in the street corner, a road junction or a deserted lane" (Chen 2017, 75).

Drawing on Tim Ingold's anthropological perspective, as articulated in *Lines: A Brief History* (2007), the mini-landscape method delves into walking as a means of interacting with the world, underscoring the interplay between movement and perception. Alexandra Lasczik et al. (2021) highlight the importance of considering groups moving side by side when walking together rather than adhering to traditional face-to-face teacher-student relationships. They examine this dynamic within arts education and emerging arts-based walking practices, suggesting that walking side by side creates space for pluralistic processes to unfold. In the mini-landscape method, when groups walk side by side instead of following a leader, it enables anyone to take the lead or share their expertise, knowledge, or feelings during the walk, thereby democratizing the research process.

This perspective aligns with Manning's (2013) concept of "more-than-human movement," which suggests that what we perceive is never simply a subject or an object but an ecology. It explores how movement and relationality transcend individual human boundaries, examining how bodies, environments, and nonhuman entities coalesce in a shared space of interaction. Manning proposes that this form of perception, experienced particularly by neurodivergent individuals, such as those with autism, forms the basis of all human perception, understood as an ecology. Yi Chen (2017) adds that movement is a form of social practice that both reflects and shapes cultural and social dynamics. This viewpoint complements the mini-landscape method by reinforcing the idea that walking together as a group is not merely a physical activity but a shared social practice. For example, figure 12.4 illustrates participants moving through the landscape as they support one another.

The group walk culminates in a prepared area featuring vivaria arranged in a semicircle, along with tools. This staging area, reminiscent of a natural amphitheater, is designed to evoke a sense of wonder without being intimidating, echoing Relph's (1976) notion of meaningful places. It serves as a reflective space for collective engagement, welcoming all participants, regardless of the paths they have taken.

Embodiment 2: Creating Mini-Landscapes

"I touch one of my lips with my middle finger. Consciousness resides in this contact. I begin to examine it. It is often hidden in the fold of tissue, lip against lip, tongue against palette, teeth touching teeth, closed eyelids . . ." (Serres 2018, 22).

Expanding on how the different bodies are interacting with the landscape, each person is now asked to work in a group to create a mini-landscape. To avoid limiting their creativity, knowledge, and feelings, participants are encouraged to create an abstract, narrative, simulated, or literal depiction of the landscape they have walked through, using materials found in the environment. This tactile engagement echoes Serres's (2018) exploration of consciousness through physical contact, encouraging the opening up of the senses.

The brief remains simple, focusing on what participants wish to express about the landscape, while researchers emphasize the importance of the group's decision-making process and how they facilitate discussions around it. Researchers also provide a briefing on what can and can't be done in the landscape, ensuring respect for both the human and nonhuman. Individuals who prefer not to collect physical materials are informed that they can instead gather other resources, such as sounds or imagery.

While working in subgroups, participants engage in tacit and tactile interactions with the landscape—touching, collecting, growing, or arranging elements. They walk away from and return to their chosen sites, gathering

Figure 12.5 Children making a mini-landscape. Source: Photograph by Andy Hughes.

materials as they go. Some create assemblages of earth and loose material, while others cultivate mini-ecosystems replicating their environment. Some groups design pools, flooded landscapes, kinetic sculptures, or earthworks. For instance, in Figure 12.5 children make a mini-landscape with mud.

Manning's (2013) work offers insights into how these mini-landscapes represent human experiences and embody the interdependent connections between people, objects, and the natural world. In this method, participants cocreate mini-landscapes that emerge from their collective interactions with each other and the environment. Manning's concept suggests that these creations are relationally "felt" landscapes, incorporating the movements and embodied perceptions of the subgroup, influenced by the nonhuman elements they engage with.

Embodiment 3: Inscribing, Laying Down, and Uttering the Cocreation

"[If] embodiment and therefore the body is taken seriously, it must also be pointed out that what can seem like even the simplest technique can provoke deep reactions and using embodied approaches might therefore give rise to

Figure 12.6 A participant adding text to a mini-landscape. Source: Permission to print approved by University of Plymouth (Copyright).

unexpected responses, even to a seemingly innocuous question like 'how are you?'" (Clughen 2024, 740).

As the second stage (Embodiment 2) draws to a natural close (approximately forty minutes), participants are asked a series of open questions. They are asked to write one or two words on slides in response to prompts such as:

- Describe what you have seen in this landscape.
- What have you heard today?
- What have you touched today?
- What smells can you describe from the landscape today?
- Can you describe any tastes from the landscape?
- How did your balance feel in this landscape?
- How did your body feel in this place today?
- How did you feel inside yourself in this landscape today?

These questions invite participants to reflect on their balance and emotions. As Lisa Clughen (2024) highlights, care must be taken to "read the space," as such questions—particularly the last three—can open up vulnerabilities. It is crucial to recognize that attempting to "separate" some senses may lead to unexpected responses. Special attention is paid to rhythmanalysis, a Lefebvrian concept, which does not differentiate between the senses but instead focuses on an overall meta-sense (Chen 2017, 2).

Mindful of potential illiteracy within the group, facilitators may encourage participants to draw on slides or use other methods to express their feelings. This approach emphasizes the act of creation and cocreation, prioritizing sensory experiences over written communication. Participants are empowered to engage as beings who sense and feel, avoiding an excessive focus on writing and documentation. However, the slides still serve to introduce a sense of scientific exploration, demonstrating that participants' words (or images) are valuable and poetic, worthy of being documented alongside their mini-landscapes. (Figure 12.6 illustrates a participant adding a slide to the group's mini-landscape.)

The group reconvenes to encourage further cocreation, with participants deciding which words or slides to add to their mini-landscape. Participants who are uncomfortable sharing specific words can choose to keep them private. The slides are arranged in various ways: some placed within the box, others layered, and some left out altogether.

Reflections

Rawness and Extraction

The mini-landscape method, a creative and participatory approach to engaging with landscapes, relies on raw materials—natural, unaltered elements. As one participant explained, the method allowed them to connect quickly with another's perception of the landscape, bypassing long interviews: "For me, it was an experience of very quickly getting to the heart of someone's heart of their idea of their landscape." This connection, born of raw materials, offers participants a sensory, tactile encounter that fosters a deep relationship with nature.

Incorporating raw materials into visual and creative methods promotes a cocreative rather than an extractive relationship with the environment, as Ingold (2011) suggests. This perspective frames natural materials not merely as tools but as active agents in the cocreation of knowledge. Daniel Wolter and Alexandra Toland (2023) note that working with materials like soil engages multiple senses, thus expanding sensory studies and enhancing landscape knowledge. Raw materials allow participants to explore not only the physical space but also their sensory and emotional connections to it, deepening both engagement and understanding.

The method also addresses some resource challenges in arts-led research, which often requires significant budgets for experimentation and data collection (Gergen 2015; Kester 2004). Though demanding, working with raw materials enhances the potential for embodied, sensory data to emerge. Many arts-based methods use indoor spaces and preproduced materials like paint, theater props, or clay. In contrast, this method involves working outdoors in open spaces, using found materials and various skills to draw analysis and ideas.

Miniaturization and Containment

Engaging with miniaturization is a mindful practice that requires a shift to a different scale, stimulating the imaginative aspects of the brain (Lawrence and Ellison 2023). While using a vivarium may not always be feasible due to limited resources, mini-landscapes can still be constructed directly on the ground or enclosed by natural elements like rocks, preserving the activity's process. When a vivarium is available, participants have the opportunity to explore multiple layers of the landscape, constructing homes for ground flora or creating abstract, three-dimensional sculptures. The container offers a multidimensional perspective, moving away from the single viewpoint of the Italian Renaissance garden. As one participant remarked, "The glass cases somehow made everything more focused and condensed. It became more of a story in itself."

Breathing and Making

A key feature of this approach is mindfulness in the process of making. Facilitators encourage long pauses for reflection, promoting contemplation alongside creation. In the act of making, participants inscribe not only physical forms into the materials but also emotional and cognitive traces. As one participant observed, "Our miniature landscapes expanded our gestural and verbal communications into inscriptions," highlighting how the creative process extends beyond physical crafting to include emotional expression. This practice aligns with Clughen's (2024) exploration of the body's relationship with emotion, where physical movements—such as breathing and writing—can influence both emotional states and creative output. The resulting inscriptions are as significant as the landscapes themselves, reflecting the interwoven relationship between mind, body, and material.

Rhythmic and Performative

Engaging the rhythms of the landscape through sensory experience deepens participants' connection to it. As one participant noted, sharing the experience of making the landscape brought awareness to elements often overlooked: "It really made me reflect on the things we are attuned to in landscapes (and take for granted) but perhaps never really verbalize." Following Lefebvre (2013) and Chen (2017), this method encourages participants to attune their bodies to natural rhythms, blurring the line between natural and social systems. Even as participants imagine the sounds of wind or birdsong, they engage in a performative act, breathing life into the landscape. Facilitators guide these processes, helping participants imagine and stage elements like "rain" through controlled actions, fostering a dynamic, performative landscape-making experience.

Messy and Soulful

In line with this performative, participatory way of working, the method generates rich, "messy" data by integrating physical, emotional, and cognitive responses. As one participant recalled, they were surprised by the spontaneity of the experience, such as how the sun seemed to align perfectly to create a shimmering sea effect. These moments of serendipity enhance the method's creative and intuitive nature. The approach values subjectivity and encourages deep emotional exploration within the landscape, allowing participants to connect their personal histories and feelings to the spaces they engage with.

Ethics and Practice

The method challenges traditional Western notions of nature and culture, collapsing dualities and fostering a more integrated, ethical relationship with the environment. As previously noted, for one participant, the mini-landscape "somehow made everything more focused and condensed," creating a narrative within the landscape. This process reimagines nature as an evolving and interconnected system rather than a fixed entity, encouraging a nonextractive relationship with the land, in line with frameworks of Indigenous knowledge (Smith 1999). Through ethical considerations, such as developing the vivarium with sustainable materials—three-dimensional printing technology using

Figure 12.7 A mini-landscape in a sustainably created vivarium. Source: Photograph by Dominica Williamson.

sustainable wood and plastic, and open-source software—and mindful, nondestructive engagement with the landscape, the method cultivates a respectful, reciprocal relationship with the environment. This sustainability is supported by the use of materials that have minimal environmental impact. Figure 12.7 depicts a sustainably produced vivarium.

Conclusion

Addressing global environmental challenges requires a holistic view. The concept of "landscape character" offers such an approach, proposing an interdisciplinary method to assess both scientific and social science aspects of a defined area. However, current government policies and language often fail to achieve the desired outcomes. There is a need to develop and practice forms of language that go beyond the quantifying specifics of current government landscape policies, which view land primarily as capital—a resource to be used and exploited for human benefit alone.

In the United Kingdom, the mini-landscape method suggests the need to reexamine the LCA and ELC and to adopt an approach that places an embodied sensorium at the heart of its relationship with the natural world. The method highlights the need for additional segments on the LCA wheel by exploring perception at a deeper level—such as "balance," "direction," and "seasonality," which are derived from the placement of slides within the vivarium, words like "wobbling," "collapse," and "end of summer," as well as comments from participants (Putt 2016). Figure 12.8 captures the mini-landscape with a slide labeled "End of

Figure 12.8 A mini-landscape with words placed in it such as "end of summer"
Source: Photograph by Andy Hughes.

Summer." However, it also points to problems with the compartmentalized nature of the LCA wheel. Rather than adding further segments, perhaps it would be more appropriate to blur the lines to better accommodate the messy nature of the data.

In a number of countries in the Global South, such as the Philippines, governance practices are often influenced by colonial legacies that fail to respect Indigenous epistemologies. Dario Novellino (2024) critiques the environmental governance in the Philippines, noting that it remains deeply rooted in colonial ideologies. He argues that these approaches are often top-down, prioritizing state interests over Indigenous perspectives.

The mini-landscape method challenges Western-centric discourse, which often adopts the practices of colonization, reducing landscapes to resources to be managed or exploited for human benefit (Smith 1999). It demonstrates how landscape governance can be more inclusive and participatory, ensuring that diverse perspectives are considered. This approach redefines governance within a relational context, challenging traditional top-down models. It fosters a bottom-up approach that integrates democratic and inclusive elements into data collection. Such data can be powerful when used as a layer within spatial planning tools, allowing planning processes to incorporate both community datasets and professional judgments. As an interdisciplinary approach, the mini-landscape method considers both physical and social science elements, leading to a rethinking of landscape assessment and management—a shift that opens the door to tackling global environmental challenges locally, using local knowledge.

Acknowledgments

Thank you to all the workshop participants and facilitators; Ryan Putt, for helping to shape the initial framework of this chapter; and Emily Höckert, Outi Rantala, and Veera Kinnunen, for their recent contributions to the method.

References

Abram, David. 1997. *The Spell of the Sensuous: Perception and Language in a More-than-Human World*. Vintage.

Chen, Yi. 2017. *Practising Rhythmanalysis: Theories and Methodologies*. Rowman & Littlefield International.

Clughen, Lisa. 2024. "'Embodiment Is the Future': What Is Embodiment and Is It the Future Paradigm for Learning and Teaching in Higher Education?" *Innovations in Education and Teaching International* 61 (4): 735–47. https://doi.org/10.1080/14703297.2023.2215226.

Council of Europe. 2000. *European Landscape Convention.* Florence 20 October 2000. https://www.worldlii.org/int/other/treaties/COETS/2000/2.html.

Fodor, Jerry A. 1983. *The Modularity of Mind.* MIT Press.

Fox, Nick J., and Pam Alldred. 2014. "New Materialist Social Inquiry: Designs, Methods and the Research-Assemblage." *International Journal of Social Research Methodology* 18 (4): 399–414. https://doi.org/10.1080/13645579.2014.921458.

Gergen, Kenneth J. 2015. *An Invitation to Social Construction.* Sage.

Gubrium, Aline, Krista Harper, and Marty Otañez, eds. 2016. *Participatory Visual and Digital Research in Action.* Routledge. https://doi.org/10.4324/9781315422978.

Hermann, Anna, Sabine Schleifer, and Thomas Wrbka. 2011. "The Concept of Ecosystem Services Regarding Landscape Research: A Review." *Living Reviews in Landscape Research* 5 (1): 5–26. https://doi.org/0.12942/lrlr-2011-1.

Ingold, Tim. 2000. *The Perception of the Environment: Essays on Livelihood, Dwelling and Skill.* Routledge.

Ingold, Tim. 2007. *Lines: A Brief History.* Routledge.

Ingold, Tim. 2011. *Being Alive: Essays on Movement, Knowledge and Description.* Routledge.

Kester, Grant H. 2004. *Conservation Pieces: Community and Communication in Modern Art.* University of California Press.

Lakoff, George, and Mark Johnson. 1999. *Philosophy in the Flesh: The Embodied Mind and Its Challenge to Western Thought.* Basic Books.

Lasczik, Alexandra, David Rousell, and Amy Cutter-Mackenzie-Knowles. 2021. "Walking as a Radical and Critical Art of Inquiry: Embodiment, Place and Entanglement." *International Journal of Education Through Art* 17 (1): 3–11. https://doi.org/10.1386/eta_00047_2.

Landscape Institute. 2016. "Technical Information Note 08/2015". *Landscape Character Assessment.* Landscape Institute. https://www.landscapeinstitute.org/wp-content/uploads/2016/01/Landscape-Character-Assessment-TIN-08_15-20160216.pdf.

Lawrence, Ceri, and Diane Ellison. 2023. "A Small-Scale Study Exploring the Efficacy of Scale Modelling as an Open Intervention for Military Veterans with Mental Health Conditions." *British Journal of Occupational Therapy* 87 (3). https://doi.org/10.1177/03080226231208053.

Lefebvre, Henri. 2013. *Rhythmanalysis: Space, Time and Everyday Life.* Translated by Stuart Elden and Gerald Moore. Bloomsbury.

Manning, Erin. 2013. *Always More Than One: Individuation's Dance.* Duke University Press.

Martin, John, Dominica Williamson, Klara Łucznik, and John Adam Guy. 2021. "Development of the My Cult-Rural Toolkit." *Sustainability* 13 (13): 7128. https://doi .org/10.3390/su13137128.

Merleau-Ponty, Maurice. 1962. *Phenomenology of Perception*. Translated by Colin Smith. Routledge and Keegan Paul.

Metzinger, Thomas. 2009. *The Ego Tunnel: The Science of the Mind and the Myth of the Self*. Basic Books.

Novellino, Dario. 2024. "'The Myth of Integrating Local Knowledge': Living and Ideological Landscapes on Palawan Island, the Philippines." *Journal of Political Ecology* 31 (1): 909–38. https://doi.org/10.2458/jpe.5968.

Pink, Sarah. 2015. *Doing Sensory Ethnography*. Sage.

Putt, Ryan. 2016. "Discovering Woodland Landscapes: A Perceptual and Economical Valuation Assessment and Comparison Between Two Sites of Similar Character in South West England." Unpublished master's thesis, University of Plymouth.

Relph, Edward. 1976. *Place and Placelessness*. Pion.

Serres, Michel. 2018. *The Five Senses: A Philosophy of Mingled Bodies*. Translated by Margaret Sankey and Peter Cowley. Bloomsbury Academic.

Smith, Linda Tuhiwai. 1999. *Decolonizing Methodologies: Research and Indigenous Peoples*. Zed Books; University of Otago Press.

Tilley, Christopher. 1994. *A Phenomenology of Landscape: Places, Paths and Monuments*. Berg.

Tudor, Christine. 2014. *An Approach to Landscape Character Assessment*. Natural England. https://assets.publishing.service.gov.uk/media/5aabd31340f0b64ab4b7576e/ landscape-character-assessment.pdf.

Williamson, Dominica, Kayla Parker, and John Martin. 2016. *Tagscape by Ecogeographer—Issuu*. https://issuu.com/ecogeographer/docs/tagscape_catalogue _spreads_lowres.

Wolter, Daniel, and Alexandra Toland. 2023. "Soil Art: Sensory and Symbolic Engagement with Soils." In *Encyclopedia of Soils in the Environment* (second edition), edited by Margaret Oliver and Michael J. Goss. Elsevier.

13

Natural Training

Exploring Historical Ideas and Potential Futures of Sustainable Running

Daniel Svensson

Running is the union of walking and flying, yet today it often seems to be more about the alliance between science and technology. The very developments that have helped raise performance levels—through scientific knowledge, better equipment, and more specialized training—have also distanced sport from nature and, according to some, alienated us from our own sensory experiences (Melin 2018).

This chapter explores the idea of natural training and, more specifically, whether it has the potential to answer some of today's pressing questions about environmental sustainability and human agency. "Natural training" is a term I use to conceptualize the historical idea that training, whenever possible, should be done using natural movements outdoors, in natural surroundings, grounded primarily in personal and experience-based knowledge, and with limited use of technology.

I will analyze natural training and its relation to the landscapes in which running is practiced. The chapter will draw on historical examples of how certain places have become important for running over time. The aim is to explore whether—and how—the tradition of natural training can be relevant in shaping the sustainable landscapes (that is, landscapes that are used without being overexploited) of tomorrow. Most examples will come from the practice of running, and this chapter aims to critically examine the balance between science and experience in training over the past century.

I draw on material from several research projects I have conducted over the last ten years, all of which have to some extent explored the balance between personal/experiential and local knowledge on the one hand and scientific/theoretical knowledge with claims of a more universal validity on the other. Based on this, I will argue that natural training and the traces it leaves are part of a "movement heritage"—a concept used to understand the remnants of human mobility on foot as an important historical legacy and a potential resource for the heritage sector. This has implications not only for the future of sport but also for the potential of sustainable development.

The Context of Running

The notion that we are affected in various ways by our geographical context has been established for a long time. In ancient Greek philosophy, the idea that climate determines the human bodily constitution and way of being was influential (Hepach 2022). There is also a long tradition of romanticizing how outdoor training strengthens both the body and mind, as well as the connection to nature and, ultimately, to the spiritual world. As the Swedish endurance sport coach and filmmaker Gösta Olander declared in the late 1940s, "Our Lord has, through a rich and varied nature, given us the best training ground" (Olander 1948, 4).

In addition to the role of nature as an arena for training, there is also a strong tradition of focusing on natural movements and exercises, as suggested by the French physical educator Georges Hébert in his *méthode naturelle* (2014 [1910]). Hébert, Olander, and many others shared a holistic perspective on training that connected physical activity to nature. To what extent is this particular holistic tradition still relevant today? In this chapter, I will argue that it holds significant potential in advancing the sustainability of sport.

In a longer historical perspective, most sports are deeply embedded in and derived from specific landscapes. Many outdoor sports emerged in places with particular conditions—we used to ski where there was snow, surf where there were waves, climb where there were mountains, and run where there were paths. To borrow a concept from Brian Wilson and Brad Millington (2016), most outdoor sports were "contextual sports." This is still true to some extent, but more than a century of scientific and technological advancement has changed the way we train. Today sport is less grounded in personal experience and local landscapes, while the impact of science, technology, and standardized indoor training has grown.

The development has been labeled a "technoscientific revolution" (Fouché 2017), taking many sports quite far from their historical contexts. If the primary goal of sports is to improve performance and set new records, then it has been a remarkable success story. Results in most sports have skyrocketed. When Swedish runner Gunder Hägg set his world record (4:06.2) for the English mile at Slottskogsvallen in Gothenburg, Sweden, on July 1, 1942, he had no way of knowing his result would be replicated and surpassed by hundreds of runners. Present-day world records are understood very differently. We have come to expect constant improvement, both on an individual and a collective level. Although Hicham El Guerrouj's world record time of 3:43.13 is a remarkable achievement, it is unlikely to stand forever.

Even running—a sport that, until a few years ago, was minimally affected by technology—has seen an approximately 25 percent increase in one-hundred-meter sprint times over the last hundred years. This underscores that more and better training influences performance levels. In sports with a strong dependence on equipment, the impact of science and technology is even clearer: the one-hour cycling record has improved by 221 percent since 1900 (Haake 2009). Cross-country skiing has evolved from a means of transportation across snow-covered lands on wood-based equipment to a high-tech sport in which the test lab plays an integral role (Svensson and Sörlin 2019).

Since the birth of modern sports, we have strived to follow the credo of the International Olympic Committee: *Citius*, *Altius*, *Fortius* (Faster, Higher, Stronger). Sports have made significant contributions to our societies in many ways and are integral to achieving some (if not all) of the United Nations' seventeen Sustainable Development Goals (Kellison et al. 2022). At the same time, however, sports also have an environmental footprint that has become increasingly problematic over time. Elite sport has been highlighted as the most challenging in this regard (Gammelsæter and Loland 2022), and it is also at this level that everyday practice has deviated the most from the natural training tradition. It is challenging to align the pursuit of increased performance with high ambitions for environmental sustainability (Loland 2001; Svensson et al. 2023).

If we view sports primarily as a quest to increase performance levels and break records, then the last hundred years have been a remarkable success story. But what happens if we also consider sport and other forms of outdoor mobility as a way to connect with nature? What do the developments of the last century mean for the potential of environmentally sustainable sport?

Over the course of the twentieth century and into the present, many sports have undergone a process of "indoorization" (Backman and Svensson 2022; Svensson et al. 2023), which has challenged a fundamental aspect of why we engage in training in the first place: namely, to move through the landscape using our own bodies. We will return to this point, but first I want to provide a historical overview of the connections between sport and nature from a Swedish perspective.

Fragments of Natural Training

Is there such a thing as "natural" training? If so, what is its essence? In this section, I will explore how it relates to various aspects of running, including training setup, equipment, landscapes, and philosophy. First, we must acknowledge the longstanding belief that being in nature is good for us. The importance of the outdoors for a healthy life has a long tradition in the Nordic countries and has been developed by many authors, scholars, and coaches. Discussions about the role of indoor versus outdoor training, and nature's impact on health and well-being, can be traced back at least to the eighteenth century (Kolbe 2024). Outdoor exercises in natural surroundings also have a long history as part of military training (Försvarsmakten 2024).

Given that Sweden has historically been a sparsely populated country, with long distances between major cities and vast areas of accessible outdoor landscapes for training and recreation, outdoor training has long been a strong tradition. However, as early as the second half of the nineteenth century, concerns emerged that training was increasingly becoming an indoor practice, to the detriment of both physical performance and individual well-being. The gymnastics instructor Anton Santesson was a vocal critic of how Swedish Ling gymnastics trained children, arguing that the emphasis on repetitive indoor sessions made gymnastics boring and even unpleasant. Instead he advocated for outdoor training and natural activities like swimming, running, and climbing (Santesson 1866). For Santesson and others, there was intrinsic value in training outdoors, being in nature, and learning to master a variety of conditions—all of which were considered integral to training. However, not everyone agreed.

The essence of the disagreement within the Swedish gymnastics movement largely revolved around issues of predictability and comparability. Was nature an asset precisely because it defied control, or was it a threat to the rational approach because it made it more difficult to control and compare the performances and outcomes of training? This debate continued throughout the twentieth century, intensifying in the 1940s following physiological research from the Swedish School of Sport and Health Sciences (GIH). This department, led by the Danish professor Eric Hohwü-Christensen, included physiologists such as Per-Olof Åstrand, Irma Åstrand, and Bengt Saltin. They began to articulate a rational training model based on scientific knowledge, in contrast to the personal, experiential knowledge that had previously been most valued.

Hohwü-Christensen sought to rationalize training practices, particularly for children and adolescents. He hoped that "different private tricks and superstition would disappear from the training ground" (Hohwü-Christensen 1943, 178). For Hohwü-Christensen, experiential and personal knowledge should be questioned because it was based on tradition rather than rational, falsifiable empirical evidence. And there were plenty of examples to critique. How, for instance, could it be considered important to drink alcohol after training, or to claim that training in a specific place was superior to any other? A scientific turn was not only necessary but long overdue, according to Hohwü-Christensen and others who shared his views.

However, when physiological research began to influence how athletes and coaches conducted their training, not everyone was impressed by the emphasis on comparability, indoor training, and laboratory tests. As Gösta Olander, at the Swedish ski and outdoor sport mountain station Vålådalen, declared in a letter to the Swedish Ski Association in 1955: "I think they get stuck too much on theoretical details, too school-like, scientific, while I mean that it is like with everything else, practice and more practice that is primarily needed so that you get a pure, emotional empathy for your sport" (Olander 1955). This quote illustrates how Olander believed that the landscapes in which we train, and our experiences of moving through them, can have profound emotional significance. Landscapes matter, both historically, as part of various training regimes, and today, as heritage and a resource for current and future training. The landscapes of training are, at least in theory, examples of "responsible inheritance" (Dooren 2014), a movement heritage (Svensson et al. 2022) in which we can see traces of the past and visions of the future. In sport (auto)biographies of endurance athletes in a Nordic context, the importance of the local landscape is consistently emphasized.

For Olander, who was inspired by the Sámi people's ability to move through the mountain landscapes, training was a holistic endeavor that could not be separated from nature. Indeed, certain types of landscapes and terrains—particularly those found in Vålådalen in the Swedish mountain range—were considered not only superior to indoor training but also superior to any other type of outdoor training ground. According to Olander, one should use the existing landscape to shape one's training, not the other way around (Olander 1948). These views stood in stark contrast to the emerging idea of rational training based on physiological research, which emphasized the scientific values of predictability, comparison, and control. Similar ideas were held not only by coaches but also by athletes. Norwegian skier Ingrid Wigernæs (1928–2023) is one example. In her autobiography, she expressed concern about monotonous training and shared that, in the toughest training sessions, "we can find joy in bird song and evening sun" (Wigernæs 1967, 94–95). The Swedish skier Nils "Mora-Nisse" Karlsson (1917–2012) also voiced his opposition to "rational training" for young athletes, advising them instead to focus on simple, natural training (Karlsson 1953, 108).

Even in a setting as structured and elite-oriented as Swedish upper-secondary sport schools for talented athletes, the tension between rational and natural training persists. While the importance of lab tests and indoor training with specialized equipment continues to grow, each school still relies heavily on its immediate surroundings. Geographical features such as hill paths or a long serpentine road up a mountain are used not only to articulate performance but also to build community and foster an understanding of the place and its history (Svensson 2024). Place still matters, for both athletes and coaches. We see a similar emotional connection to the landscape and a comparable determination to protect it among ski instructors in alpine skiing (Larneby 2024). This is partly based on an instrumental understanding of sporting landscapes—without them, athletes cannot train as they would like to. Studies of outdoor sports practitioners in Sweden have indicated that the environmental dimension, including the values of being in nature, is important to athletes, but that they tend to focus more on how well the training landscape supports their performance (Backman and Svensson 2022).

There is, however, also a more environmental dimension, expressed as a determination to safeguard natural training landscapes for future generations. It is not surprising to see ideas like these expressed by winter sports enthusiasts and professionals, as these sports are perhaps the most visibly affected by environmental degradation and climate change in the Northern Hemisphere. The deeply emotional plea to "protect our winters" is just one of many examples of how athletes are using their voices to challenge the unsustainable aspects of modern sports. As American skier Jessie Diggins put it, "I'm speaking up about how much winter means to me because I want it to stick around. I want my grandchildren someday to be able to try cross country skiing" (Diggins n.d.).

Even some of the architects behind a more scientific approach to endurance training—physiologists from the Swedish School of Sport and Health Sciences in Stockholm—have advocated for a more mindful approach toward places for recreational walking and running (Svensson & Sörlin 2019). To name just one example, Peter Schantz has written for decades about the need to protect nature in the context of sports (one early example is Schantz 1991). Other sports physiologists, such as Arthur Forsberg and Hans-Christer Holmberg, have expressed concern about the risks of overexploiting nature. The difference between natural and rational training does not lie primarily in the valuation of nature but in the balance between personal experience and scientific data.

The historical development of running has, unsurprisingly, moved toward more advanced technology, higher speeds, and improved performance. The natural practice of running barefoot was largely abandoned but has experienced a revival in recent years. A desire for a simpler way of running, along with a reaction to Bill Bowerman's development of a new style of running in the 1960s and 1970s—and the subsequent need for more advanced shoes to support this heel-driven technique (McDougall 2009)—sparked disagreements with the running coach Arthur Lydiard. Although Lydiard coached elite athletes, he valued the simplicity of running and the natural ability to run, as seen in children: "Kids have been running through the centuries. They mostly run barefooted so their feet develop properly and naturally" (Lydiard and Gilmour 2011, 23). This appreciation of natural movements in natural surroundings is quite similar to the ideas of Anton Santesson in the mid-1900s.

Today we see how such ideas are linked to environmental sustainability. When Swedish endurance event organizer Gael Joly sought to create a sustainable trail running event, he focused on the responsibility of all involved to care for the event's location, a nature reserve in Skåne, southern Sweden (Österlen Spring Trail 2024). He is not alone but part of a growing movement advocating for more sustainable running practices. The Kilian Jornet Foundation, named after the Spanish trail runner, is also addressing similar issues and is involved in the European project Green Trail Concept (Green Trail Concept 2024). These are just a few examples of how environmental sustainability is challenging and reforming endurance running. Though these initiatives are new, they are rooted in longstanding traditions, linking them to figures like Hébert, Santesson, Lydiard, and Olander.

The Path Ahead

As we run along the paths leading through time, we must keep in mind the fundamental question: Why? Why do we run? Are we running from something—aging, BMI, anxiety, our past? Or are we running toward something—endurance, freedom, eternity? Probably both, and countless other reasons. Depending on the answer, science and technology can help open new avenues for motivation, provide data to guide our pacing, highlight progress, and connect us with others on the same path.

However, technology can also obscure. It can cause stress and separate us from each other and ourselves by measuring exact distances. And when it does, it is not acting on its own. Technology does not (yet) have agency. But we do, when we turn sports into a tech-dependent, science-based, and data-driven endeavor, and when we imagine and create different futures where GPS watches, heart rate monitors, and AI-powered training programs complement, rather than replace, our experiential knowledge.

In November 2024, I was a guest professor at the University of Turin. On a calm and sunny day, I left my apartment at the corner of Via Andrea Doria and Via San Francesco da Paola. I ran through the old streets of the city, with its echoes of Roman Empire urban planning, down to the river Po, across the bridge, up the hill past Monte dei Cappuccini, and further up to the Parco di Villa Genero. Every step was recorded. Yet if I were to look at the data twenty years from now—speed, watts, cadence, pulse, elevation, and distance—it would not tell me anything useful. The sunlight reflecting on the Po, the smell of coffee from a small café, the laughter of children in the schoolyard on the hillside, or the squirrel outrunning me on the path to the top—these are the memories that will stay with me. This is why natural training is enriching, contributing deeper values in a time dominated by technological monitoring.

This is not about conservatism, nor a Luddite rejection of new technology. Rather, it is about not abdicating our position as beings with the capacity to improvise. Had the battery of my GPS watch died, I might have questioned whether the run even happened. If it's not recorded, is it actually real? Yes, because I was there. So were the river, the trees, the birds, and the squirrel. They were the context of my contextual sport, and they are not dependent on any technological mediation to have a lasting—I am tempted to say eternal—place in my memory. I doubt I had any similar impact on the existence of the river, the trees, the birds, and the squirrel. Yet if they do somehow remember me as part of their context, hopefully it is only as a fellow creature temporarily passing by, using our shared potential for mobility.

What does this have to do with movement heritage and sustainable sport? I argue that the core idea of natural training is freedom—leaving room for the unexpected, the expression of the moment, and the desire not only to improve but also to improvise. When used cautiously, science and technology can serve as a tool in this process. However, we also need to nurture our ability to manage without sensors and smartphones. By following paths—rather than relying on maps or GPS signals—we cocreate a movement heritage that can inspire others. Such a heritage can be a powerful tool on many levels (social community, public health, sustainable landscape use, and more), as researchers from many different geographical and cultural contexts have demonstrated (for example, Svensson et al. 2022; Hickman and O'Hara 2022; Österlund-Pötzsch 2022; Andersson 2024).

Sport and physical activity are important tools for achieving many of the Sustainable Development Goals. Good health and well-being are probably the most obvious, but sport can also have a positive impact on gender equality, reduced inequalities, sustainable cities and communities, and several other Sustainable Development Goals. Through theories of learning and pedagogy, we have come to understand the importance of trying and exploring. While John Dewey may never have said "learning by doing," he did emphasize the value of experiential knowledge. You can read every manual in the world about running, yet it won't bring you one step closer to discovering its joy. It is only when you learn by doing that you gain the ability to understand those text-based sources and draw from them. We live in a time and place where many of our physical abilities are not put to the test, making running, walking, and other forms of physical exercise all the more important. As Norwegian sport philosopher Sigmund Loland has stated, "In a modern, high-technological world in which everyday movement has become redundant, sports stand out with their emphasis on bodily capacities and effort" (2023, 28).

Practical knowledge is key to theoretical knowledge, while the opposite is rarely true (at least in sports). Running is not like assembling a piece of Swedish mass-produced furniture. Forget the manual and give it a try. Others have run before you, and the paths you follow are their paintings on a cave wall. While

Tomas Tranströmer may not have been a runner, he was a brilliant poet. In his poem "From March 1979," he encounters tracks in the snow and reflects on their significance: "I come upon the tracks of deer in the snow. Language but no words" (Tranströmer 2006, translated by Fulton). These tracks, left by us and others in the snow and on the ground, are signs of movement, of our ability to use our bodies in a sustainable way. They form a potential heritage of mobility, one that could strengthen our connections with both past and future generations based on our shared desire to move.

Science and technology have been integral to our understanding of what it means to be inhabitants of the Earth. But to understand what it means to be alive, we need personal experience. I have used many words here to argue for the importance of the silent track in the running experience. We need footprints in the snow—from humans, animals, and birds. We are one—through running, which is walking and flying combined.

References

Andersson, Dan. 2024. *Exploring Perceptions of Route Environments in Relation to Walking*. Swedish School of Sport and Health Sciences.

Backman, Erik, and Daniel Svensson. 2022. "Where Does Environmental Sustainability Fit in the Changing Landscapes of Outdoor Sports? An Analysis of Logics of Practice in Artificial Sport Landscapes." *Sport, Education and Society* 28 (6): 727–40.

Diggins, Jessie. n.d. "With a Medal in My Pocket." Accessed December 10, 2024. https://protectourwinters.org/with-a-medal-in-my-pocket.

Dooren, Thom van. 2014. *Flight Ways: Life and Loss at the Edge of Extinction*. Columbia University Press.

Fouché, R. 2017. *Game changer: the technoscientific revolution in sports*. Johns Hopkins University Press.

Försvarsmakten. 2024. https://www.forsvarsmakten.se/sv/information-och-fakta/var-historia/artiklar/idrotten-och-forsvaret.

Gammelsæter, H., and S. Loland. 2022. "Code Red for Elite Sport. A Critique of Sustainability in Elite Sport and a Tentative Reform Programme." *European Sport Management Quarterly* 23 (1): 104–24.

Green Trail Concept. 2024. https://greentrailconcept.eu/.

Haake, Steve J. 2009. "The Impact of Technology on Sporting Performance in Olympic Sports." *Journal of Sports Sciences* 27 (13): 1421–31. https://doi.org/10.1080/02640410903062019.

Hébert, Georges. 2014 [1910]. *The Natural Method: Georges Hébert's Practical Guide to Physical Education*. CreateSpace Independent Publishing Platform.

Hepach, Maximilian Gregor. 2022. "Ephemeral Climates: Plato's Geographic Myths and the Phenomenological Nature of Climate and Its Changes." *Journal of Historical Geography* 78: 139–48.

Hickman, C., and G. O'Hara. 2022. "Delineating the Landscape: Planning, Mapping and the Historic Imaginings of Rights of Way in Twentieth Century England and Wales." In *Pathways: Exploring the Routes of a Movement Heritage*, edited by D. Svensson, K. Saltzman, and S. Sörlin. White Horse Press.

Hohwü-Christensen, Erik.1943. "Förslag till ny utbildningsplan för lärare i gymnastik, lek och idrott." *Tidskrift i gymnastik* 10: 165–81.

Melin, Roger. 2018. "Technology and the Existential Element in Sport." *Scandinavian Sport Studies Forum/idrottsforum.org*, December 6, 2018.

Millington, B., and B. Wilson. 2016. *The Greening of Golf: Sport, Globalization and the Environment*. Manchester University Press.

Karlsson, Nils. 1953. *I vita spår*. Bonnier.

Kellison, T., B. P. McCullough, and E. N. Melton, eds. 2022. *The Routledge Handbook of Sport and Sustainable Development*. Routledge.

Kolbe, Wiebke. 2024. "Kurorte und Bäder in Westeuropa und Skandinavien vom 18. bis zum 20. Jahrhundert." In *Kurorte in der Region: Gesellschaftliche Praxis, kulturelle Repräsentationen und Gesundheitskonzepte vom 18. bis zum 21. Jahrhundert*, edited by L. Seegers, M. Frese, and M. Thießen. Wallstein Verlag.

Larneby, Marie. 2024. "'I Live with and By Nature': Swedish Alpine Skiers Reflect on Professional and Lifestyle Skiing, Nature, and Snow, 1964–2023." *Sport History Review* 55 (1): 65–86.

Loland, S. 2001. "Record Sports: An Ecological Critique and a Reconstruction." *Journal of the Philosophy of Sport* 28 (2): 127–39.

Loland, Sigmund. 2023. "The Idea of Natural Athletic Performance: An Interpretation and a Defence." In *Sport, Performance and Sustainability*, edited by D. Svensson, E. Backman, S. Hedenborg, and S. Sörlin. Routledge.

Lydiard, A., and G. Gilmour. 2011. *Running to the Top* (third edition). Meyer & Meyer Sport.

McDougall, Christopher. 2009. *Born to run: the hidden tribe, the ultra-runners, and the greatest race the world have ever seen*. Profile Books.

Olander, Gösta. 1948. *Träningsråd för skidåkare*. Swedish Ski Association.

Olander, Gösta. 1955. Letter to Sigge Bergman. Swedish Ski Association archives, documents from Sigge Bergman, 1952–1958, Swedish National Archives, Arninge, Stockholm. My translation from the Swedish.

Österlen Spring Trail. 2024. https://www.osterlentrail.se/sustainability.

Österlund-Pötzsch, Susanne. 2022. "Archipelagic Paths: Narratives, Heritage and Community in Public Trail Walking on the Åland Islands." In *Pathways: Exploring the Routes of a Movement Heritage*, edited by D. Svensson, K. Saltzman, and S. Sörlin, 131–51. White Horse Press.

Santesson, Anton B. 1866. *Skol-gymnastik för rikets elementar-läroverk jemte en kort skildring af disciplinens tillstånd vid flera af dessa läroverk. 2. tillökta uppl.* Bonnier.

Schantz, Peter. 1991. "Idrott och miljö." *Tidskrift i Gymnastik & Idrott* 6: 41–46.

Svensson, Daniel. 2024. "Landscapes of Performance: Using Local Geography for the Testing of Sport School Pupils in Sweden, 1972–2023." *Sport History Review* 55 (1): 31–47.

Svensson, Daniel, and Sverker Sörlin. 2019. "The 'Physiologization' of Skiing: The Lab as an Obligatory Passage Point for Elite Athletes?" *Sport in Society* 22 (9): 1574–88.

Svensson, D., K. Saltzman, and S. Sörlin, eds. 2022. *Pathways: Exploring the Routes of a Movement Heritage*. White Horse Press.

Svensson, D., E. Backman, S. Hedenborg, and S. Sörlin, eds. 2023. *Sport, Performance and Sustainability*. Routledge.

Tranströmer, Tomas. 2006. *The Great Enigma: New Collected Poems*. Translated from the Swedish by Robin Fulton. New Directions.

Wigernæs, Ingrid. 1967. *Mot mål med jentutn*. Aschehoug.

14

Telematic Soundwalking

Toward the Sonorous Step

Maria Papadomanolaki

In 2010, during a stay in New York, I came across a poem by Allen Ginsberg titled "Manhattan Thirties Flash" (Ginsberg 1973). The poem carries a process of listening, walking, thinking, and writing, trying to keep up within the belly of the city of Manhattan:

> Long stone streets inanimate, repetitive machine Crash
> cookie-cutting
> dynamo rows of soulless replica Similitudes brooding tank-
> like in Army Depots
> Exactly the same exactly the same exactly the same with no
> purpose but grimness . . .

The writer's pace assumes the sensing body of a walker inside a landscape. Within its short length, the poem manages to encapsulate the essence of how it can feel to be part of this big city and to glide through its busy streets. Most importantly, the moment I read it, I immediately began to breathe in my inherent and embodied experiences and found myself critically engaged with Ginsberg's vision. An important exchange was established between his poetic voice and my own. The text functioned as a key trigger, allowing my inner processes to seep through, be sounded out, and significantly affect the way I connect to the city where I live and walk, while helping me to resist the tendency to normalize urban environments as "pass-through" and impersonal spaces.

Similarly to Ginsberg's comment on a city in a state of mechanistic metamorphosis and regulated functionality, a pass-through space can be defined as a site that we pass through, either for commuting or temporary leisure. The poem, in contrast to the transient and regulated pace of urban spaces, caused me to slow my steps, reflect on the sounds of the city, and recall past encounters with these sounds and how they become carriers of the intricacies of an environment.

My experience and reading of Ginsberg's poem made me reflect on the possible ways to break the self-centered perceptual bubble often adopted by city dwellers and facilitated by iPod playlists and noise-canceling strategies, and to open up similar pass-through spaces within the city to be reinvented, sensed, and subverted. It also led me, by setting aside any notions of skilled and conscious listening, to want to highlight the importance of allowing our bodies to be heard and felt in our day-to-day walks in these spaces. A shift has therefore been initiated, prompted by my reflection on Ginsberg's poem, toward the polyphonic nature of my own sonorous step as it cohabits within the city and leaves its mark as it passes through the city's busy train and subway stations, urban parks, alleyways, and junctions.

Walking and Voicing as Listening

Listening from within, as Tim Ingold suggests, calls for a process of participating in the urban environment (Ingold 2000, 244). Such a perspective forms part of a multitude of theories, terms, and definitions regarding listening that range from the semantic aspects of meaning (semantic listening) (Chion 1994), to its emotional undertones (affective listening) (Gallagher 2013; Simpson 2009), and its power to trigger associations (associative listening) (LaBelle 2010). In the context of walking and participating from within, listening as part of a meshwork of interactions may encompass all of the above. Most of the time, what is the most important is to allow oneself to connect to an environment through experiencing, where "listening" is not necessarily limited to "listening to the space" and its acoustic and material aspects but also listening to oneself, holistically and in simultaneity with the environment. This perspective is often explored through soundwalking or simply by walking and actively listening. Beyond specialist knowledge of trends on how to (accurately or perfectly) listen, I would argue that everyone possesses the inherent ability to listen in this way, provided it is activated and brought to the foreground. That is, at least, an important hypothesis often reflected in my own work with telematic soundwalking, as I will discuss later.

This approach shifts the search for awareness toward self-reflection, listening to the environment by simultaneously listening to the self and the outside world, and toward others. Tim Ingold's concept of "taskscape" emphasizes the importance of collaboration within a community for the performance of everyday tasks, often manifested through sound. The use of movement and speech, along with the sounds of the environment, are all influenced by the temporality of the landscape, changing as the seasons change (Ingold 1993, 170–72). The shifting and subjective, the bodily and collaborative, all become central in this effort to inhabit and reinvent an environment from within. As Brandon LaBelle highlights in reference to the Situationist International (SI), it is "intuitive and spontaneous actions (such as the dérive, or drift) that underscored the [sonorous] body as a differentiating break onto the mechanistic functionality of modern life" (LaBelle 2004, 215). It is through these "down below" encounters of "the ordinary practitioner" (the walker), as de Certeau writes, that pass-through spaces can be reinvented (de Certeau 1984, 93, 98). The methodology described later takes this understanding as a starting point to create a series of performances, where the self becomes extended by attending, through walking, to other voices, people, sounds, and environments.

Listening, walking, thinking, and talking often coexist as forms of "a participatory communion of self and other through shared immersion" (Ingold 2000, 247). The thinking and speaking voice becomes an instinctive way of attuning to the sociocultural and experiential implications of a space (Pinker 1994, 17)—a space that becomes animated through walking. Don Ihde writes that "[i]n speaking, the voice 'sounds through' from the inside to the outside; in hearing it conversely penetrates from the outside to the inside" (2007, 247). In my telematic soundwalking practice, by actively participating in these exchanges of intersubjectivities, the voice—whether speaking or thinking—can significantly assist in merging internal and personal perceptive cues with external stimuli. This idea is also discussed in detail from a phenomenological perspective by Maurice Merleau-Ponty (1962). Walking, voicing, and listening therefore become crucial for creating spontaneity and intimacy within pass-through spaces (Lefebvre 1991, 70).

Listening to the Atmosphere

The relational dynamics between a particular space and how subjective and spontaneous exchanges are manifested within it are related to how the space is perceived. In this context, the concept of atmosphere—an invisible, shadow-like presence in a space—assumes an important role, especially when considered from the perspective of sound. Gernot Böhme discusses sound as one of the many layers that compose the atmosphere of a space, where its elusive and ephemeral spatial movement modifies the bodily experience of the space by adding emotional undertones. According to Böhme, atmospheres are phenomena experienced through antitheses that affect and change our mood, acting as the gluing matter between the materialities of a space and human sensibilities and senses (Böhme 1993). Extending this idea, LaBelle argues that an atmosphere can exist both independently and in conjunction with the physical attributes of a space, both in harmony with (as a binding force) and as an interrupter of the narrative of a particular environment (LaBelle 2004, 216). In this sense, staging or affecting the atmosphere of a space can become, as LaBelle suggests, a very useful strategy for reactivating our engagement with the pass-through sites by countering existing narratives and retuning our thoughts, senses, and bodies to the particularities of a site. This consideration is a key aspect of the methodology I will discuss in the following section.

A Methodology for Opening Up

A Certain Geography[1] is as much a project as it is my methodology for opening up and experiencing pass-through spaces. It is my way of addressing and engaging with all the ideas and concepts analyzed earlier. How can I trigger and refocus our inherent abilities to listen from within, together, in communion? In what ways can walking and voicing create spontaneous and intuitive sonic exchanges? How can a staged atmosphere and conditioned experience act as interrupters to a particular narrative of a space, opening it up to allow for more intimate exchanges?

The project is a continuation of a series I carried out between 2007 and 2008, titled *Trajectory*, which explored (sound)walking, colistening, and the use of networked live mobile sound transmission. As in many examples of telematic art—particularly telematic music and sound art—the emphasis is on breaking down the dynamics (situated or other) between the center and the periphery (Lemmon 2019, 5), as well as on creating spaces of intimacy and cocreation (Oliveros 2009). Telematic art prioritizes the output of an exchange over the artwork itself (Ascott 2003, 233).

In *A Certain Geography*, I further develop my practice within this line of work by introducing a two-part methodology. The first part involves performing repetitive walks over a short period within a specific site to choose the path for a soundwalk. During the walks, I interact with the site, initiate conversations with people, and take notes on sounds, my walking, thoughts, events, and fragments. This process leads to the creation of a locative text-score, which is then used in the second part of the methodology: the performance.

The text-score often functions as a textual layer that, much like the effect of Ginsberg's poem on me, creates a sense of disruption by opening up the atmosphere of the listening space. The text-score can take various forms (print or video projection), depending on the space and context. It is the manifestation of my walking body and of sociopolitical, material, or historical research. It often serves as a map for my walks. I frequently record myself reading parts of the text-score on a Dictaphone, which I also use during the performance as an additional listening layer that adds to the atmosphere of the work.

The second part of the methodology involves a site-specific performative piece. The performance consists of me transmitting my live soundwalk via the internet back to the chosen site, where the audience is located, listening. In some cases, the performance may include sharing the text-score with the audience by projecting or printing it. In other cases, it may involve other artists spontaneously improvising with me in different ways. The transmission loop is completed when I return to the audience site. Where possible, the audience is invited to participate in the exploration by sending me messages via their mobile phones—asking questions or encouraging me to take actions or deviate from my course—or by responding in any other way they see fit.

A Certain Geography is a project whose iterative and interwoven layers create an intimate and delicate space for spontaneous and intuitive exchanges, where the individual and the collective exist contingently. Taking as a point of departure the intent to disrupt the spatial confines of a performance space, the project generates an atmosphere that subverts the existing narrative of the space and enriches our listening experience within the identified space by weaving together a meshwork of simultaneous remote and local sonic layers.

Walking becomes the plateau for initiating dialogue and using the body/voice axis as a lens through which to sense ongoing interactions. The geography of both the space and the remote site, with all their underlying intricacies and ambiguities, are reciprocally contested, opened up, and reinvented (de Certeau

1984, 93, 97, 99) through the perpetual creation of a walking trajectory. Words, gestures, and sounds are repeated, countered, disjointed, and destabilized by silences, delays, and disruptions of signals.

A *Certain Geography*: Bethnal Green

In this section, I will discuss a presentation of this project as part of the "Forms of Ventriloquism" event, which I curated at IMT Gallery in July 2014. The starting point for this instance of the project was a series of listening and walking workshops I conducted with students in 2013. Walking slowly in small groups of five, we discussed the architectural and sociocultural design of shopping malls and train stations in the area. At the end of each walk, each participant created a map of the experience, highlighting their main impressions of the walk.

In one of these responses (figure 14.1), the student encapsulated, on a single page, the entire spectrum of embodied and sonorous acts undertaken by the group. Each bubble on the response represents a step taken by the group. The walking became, to refer back to de Certeau, a way of activating and reinventing

Figure 14.1 The anonymised creative response used for *A Certain Geography* in Bethnal Green. Source: Photo by Maria Papadomanolaki.

these spaces. Our walking process was inseparable from the sounds it produced and prompted. These sounds, particular to each space we visited, were collaboratively enriched, contested, and experienced through walking, listening, and conversing.

During the first part of my preparation for the performance, I felt that this particular creative response (figure 14.1) strongly resonated with my own embodied knowledge of the area. I used the response's words as stepping stones for my own repetitive walks, focusing on shops and passageways where quiet overlaps with the periodic noise of trains, where multiple languages can be heard, and where steps and movement can be amplified. Prompted by the word "Differences" found in the creative response, I walked on different surfaces—cement, mud, grass—and also through neighborhoods undergoing gentrification and those that remained "scruffy" and old. All these exchanges were then incorporated into a new text-score that I would use in the performance. The text-score acts as an intimate zone where I weave in the student's words, the sounds of thoughts and of the city, and ultimately the city itself. In its fabric, I "sound out" many of the key observations from my research, such as the importance of movement (and walking), my thoughts on atmospheres, and the significance of becoming part of the polyphonic city.

For the performance, I invited British sound artist Robin Rimbaud (aka Scanner), a local artist in the area at the time. Rimbaud had worked extensively with improvising with live sound sources, such as police scanners and mobile phone conversations, bringing "into focus issues of privacy and the dichotomy between the public and the private spectrum . . . capturing and highlighting the threads of desire and interior narratives woven into everyday life" (Joseph-Hunter et al. 2011, 41). He was more than happy to improvise inside the gallery to the live transmission of my soundwalk, which was taking place outside.

Aspects of the repetitive walks were reperformed during the performance itself. Walking at variable paces along passageways with echo and reverberation amplified my steps and movements. I focused on the aural and material nuances of the trajectory by emphasizing their differences. I performed by hitting surfaces (Figure 14.2 "Audio Clip 1," 05:29), improvising with railings, standing still, and matching words such as "trains," "gray," "voices," "Tesco," "movement," "walking," and "sounds" with tactile embodied engagements, footsteps, and slow and fast movements.

In an effort to merge the text-score with these tactile mobile interactions, I recited the text-score at different locations during the soundwalk and recorded

Figure 14.2 "Audio Clip 1" is a recording of the performance that was part of the "Forms of Ventriloquism" event at IMT Gallery in July 2014. To listen, please scan the QR code. Source: Photo by Maria Papadomanolaki.

myself on a Dictaphone. I made this choice because (a) the device's portability did not disrupt my walking and fit with my existing streaming toolkit, which consisted of my mobile phone, a microphone adapter, and my in-ear binaural microphones; and (b) it added an uncanny texture to my voice, making it resemble the ghostly voices in Scanner's previous work. The prerecorded text-score functioned as a narrative tool that affected my walking, prompting me to start and stop at different points during the performance.

The final performance comprised the incoming streamed sound, the immediate sound in the gallery, Scanner's improvisation, and the audience's intimate listening. The transmission loop was closed by inviting the audience to complete the text-score by writing resonant words prompted by this listening exchange at the entrance to the gallery (Figure 14.2 "Audio Clip 1," 21:27)—their response to the experience, their thoughts or feelings.

The performance lasted for twenty-eight minutes. From the audience's responses to a questionnaire circulated after the event, it is clear that the performance succeeded in creating a sonic atmosphere where everyone was attentive and engaged. The gallery director reported that many people were very thoughtful; some took notes while others seemed to be fully absorbed in the piece. The presence of the voice added an atmosphere of intimacy. As one member of the audience wrote in their assessment:

> I experienced the nature of the spoken word very close to the way that I experience the city, due to the 3D element of being in a space and perceiving many different, simultaneous, layered narratives from other, external spaces, and always perceiving it in my own way in relation to my internal narratives and associations of sounds. (Notes from the performance of *A Certain Geography*, 2014)

What is striking about this statement is that it reaffirms her active engagement with the piece by triggering her inherent ability to listen, and how her internal voice is embedded reciprocally within the polyphony of the city—an argument that aligns strongly with the methodology of this project. Another audience member emphasized the nature of the spoken word in his assessment, particularly in relation to the question about the voice, stating:

> One of the levels of the performance was of course the addition of the spoken words. For me it was significant that there was no theatricality and that they were spoken in a "natural" tone. That created a more surreal atmosphere. Their contribution was essential, otherwise the material provided by Scanner would have been weak. (Notes from the performance of *A Certain Geography*, 2014)

The voice, therefore, seemed to provide a useful connecting thread that tied together the various elements of the performance, adding dynamics and depth both to the audience's perception and to Scanner's contribution. In response to the question about the effectiveness of the technology and Scanner's role, another audience member wrote that

> [it] provided a strikingly direct superimposition of audio within the immediately observed performance. It did not add an additional reality, but an additional actor in the performance, someone who was registering a different sensibility than the

> one channelled by Scanner and doing so in real time. I was not concerned by the inside/outside dichotomy, but by the "otherness" of the experience. (Notes from the performance of *A Certain Geography*, 2014)

The fact that the listener was not "concerned by the inside/outside dichotomy" confirms the successful creation of a subversive atmosphere that permeated multiple layers of sound: the internal, the immediate external, and the streamed sounds. In addition, there was the layer created by the audience itself and its participation in completing the score.

In the evaluation questionnaire, the overall performance was described as "memorable," "unforgettable," "dreamy," "original and challenging," "intense, very present, embodied, honest, experiential, and clear." People's anticipations and their listenership were significantly enriched by this interplay of live and manipulated sound, as it created—as stated earlier—a space of ambiguity where the division between inside and outside did not matter. On the contrary, we were all participating in this ongoing meshwork of exchanges activated by my walking and voicing, through resonant words and how they coexisted, in absentia, with other sounds, materials, and voices. As one audience member said to me, "You were a ghost"—perhaps a ghost bringing to attention, through sound, the porosities and spatialities of the city as "an immense social experience that is, to be sure, broken up into countless tiny deportations (displacements and walks), compensated for by the relationships and intersections of these exoduses that intertwine and create an urban fabric" (de Certeau 1984, 103).

A *Certain Geography*: Ancient Messene

A later iteration of the project was presented in response to the city of Ancient Messene in the Peloponnese. Beginning by tracing the water path that ran from the northern to the southern part of this ancient city, which holds great historical and religious significance, I conducted repetitive walks across its various sites. These walks were further supported by historical research and conversations with local experts and other artists. Each walk focused on exploring the geographical terrain in relation to its sociopolitical and historical intricacies, and the resulting trajectory and text-score amplified the findings of the site-specific research.

My performance of *A Certain Geography* as part of "Tuned City—Ancient Messene"[2] was a difficult undertaking due to the nature of site, which is remote and exposed.[3] Because of limited signal reception, the type of theater, and the time of my performance (early afternoon), direct remote audience participation via text messaging was impossible and therefore excluded.

The resulting walk followed various points of interest—referred to as "points of action"—across the premeditated trajectory. Each point of action had a dedicated section in my text-score and indicated different ways of walking and interacting with the environment. These included, for example, using heavy steps to emphasize the acoustics of a space, kneeling slowly to get closer to the sounds of water, walking differently on different terrains (figure 14.3), and pausing to listen to the cicadas and sound installations across the site. The enduring nature of the walk and the sound of my moving body were always present in the transmitted soundwalk. My occasionally heavy breathing and the continuous sounds of steps across pebbles, stones, dirt, grass, and marble gave the performance a very strong rhythm as I approached the theater. I was accompanied by Professor Eric Lewis of McGill University, who performed brief trumpet improvisations from a distance to activate the specific acoustic properties of the spaces we passed through and to add to the atmosphere of the soundwalk.

Figure 14.3 Sonorous steps. Footsteps on different terrains. Source: Photo by Maria Papadomanolaki.

At each point in the walk, readings from my text-score were played back on my Dictaphone. These readings referred to relevant historical details, fictionalized in the text-score as spoken by a female inhabitant of the ancient city. The Dictaphone was strategically placed within the soundscape to coexist with other sounds occurring there at the time, as well as my own performative actions and steps.

The overlap of direct and remote sounds made the listening experience quite intense. As some members of the audience later explained to me, the powerful sense of proximity conveyed by the transmitted sound made them want to start walking in search of me. The performance concluded with my entry into the theater, where I gave the final part of my text-score to a member of the audience to perform live.

The performance was received with great enthusiasm and was described as meditative and surprising. This iteration of *A Certain Geography* created an ephemeral walking trajectory, where the performer's sonorous steps, bodily presence, and prerecorded voice merged with the spaces encountered, the multitude of permanent and perpetual sounds, and their histories, textures, and character. For some, this experience of listening without seeing echoed a kind of acousmatic listening, in which the source is hidden from the listener (Schaeffer 2017, 95–96). Some audience members referred to their listening as a very detailed yet somewhat opaque process due to how the on-site, online, and acousmatic sounds blended. My sonorous steps and prerecorded voice created a dramatic antithesis that some people greatly enjoyed in the performance. The soundwalk prompted the audience to listen carefully and critically in order to recognize and relate to the sounds and to feel "entranced with the experience of a journey" (Truax 2002).

Conclusion

A Certain Geography, as a methodology for opening up and reinventing pass-through spaces, allows for a certain form of collaborative and intersubjective deformation and recreation of space through walking, voicing, and listening. The attentive use of text and voice; the dynamics created between on-site/off-site, online/live, present/past, remote/direct; the conditioned listening of the audience; and the roaming body of the performer all compose a subversive atmosphere that, through sound and movement, triggers an intimate, spontaneous, and memorable experience of the space. The performer's body and footsteps become those of the audience, while her voice acts as a polyphonic canvas for critical listening, reflection, and response. As Fanny Chiarello writes in reference to *A Certain Geography*, "[by] becoming a test subject of her own methods under the gaze of spectators, [the artist] invites them to question the very nature of the separation and interaction between composer, performer, and audience" (Chiarello 2024, 205).

Walking offers the possibility of embodying our interactions with the environments we inhabit. Our sonorous steps become inseparable from our presence in the spaces in which we live and move. *A Certain Geography* serves as a reminder of our ephemeral, embodied movements through life and invites us to collaboratively extend our perceptual antennae toward that which we seem to forget or pass by. The remotely sensed sonorous steps, similarly to my reading of Ginsberg, become triggers for unlocking a listener's sensory and embodied memories of the sounds and environments that surround them.

Notes

1 *A Certain Geography*has been presented at the following events: "Sensingsite," Parasol Unit, London, April 2014; "Forms of Ventriloquism," IMT Gallery, in collaboration with Scanner, London, July 2014; "In this neck of the woods,, CSM, London, June 2015; "ISEA Disruption," Vancouver, August 2015; "Mobile Audio Fest," Marseille, November 2015; "Sound Acts," Athens, May 2016; "Sound & Environment," University of Hull, June 2017; "Borderline Festival," Athens, 2018; "Tuned City-Ancient Messene," Messene, Greece, 2018; "Resonant Frequencies Summit," DRAF, Goldsmiths, London, 2019.

2 For more information, see: http://www.tunedcity.net/?page_id=5013 and https://www.onassis.org/whats-on/tuned-city.

3 See https://www.youtube.com/watch?v=R9WIJEffMHQ to watch a short video documentation, with excerpts from *A Certain Geography* at 0:44 and 0:58.

References

Ascott, Roy. 2003. "Is There Love in the Telematic Embrace?" In *Telematic Embrace: Visionary Theories of Art, Technology, and Consciousness*, edited by Edward A. Shanken. University of California Press.

Böhme, Gernot. 1993. "Atmosphere as the Fundamental Concept of a New Aesthetics." *Thesis Eleven* 36: 113–26.

Chiarello, Fanny. 2024. *Basta Now: Women, Trans & Non-Binary in Experimental Music.* Permanent Draft.

Chion, Michel. 1994. *Audio-Vision: Sound on Screen.* Edited and translated by Claudia Gorbman. Columbia University Press.

de Certeau, Michel. 1984. *The Practice of Everyday Life.* Translated by Steven Rendall. University of California Press.

Gallagher, Michael. 2013. "Listening, Meaning and Power." In *On Listening,* edited by Angus Carlyle and Cathy Lane. Uniformbooks.

Ginsberg, Allen. 1973. *The Fall of America: Poems of These States.* City Lights Publishers.

Ihde, Don. 2007. *Listening and Voice: Phenomenologies of Sound.* State University of New York Press.

Ingold, Tim. 1993. "The Temporality of the Landscape." *World Archaeology* 25 (2): 152–74.

Ingold, Tim. 2000. *The Perception of the Environment: Essays on Livelihood, Dwelling and Skill.* Routledge.

Joseph-Hunter, Galen, Penny Duff, and Maria Papadomanolaki. 2011. *Transmission Arts: Artists & Airwaves.* PAJ Publications.

LaBelle, Brandon. 2004. *Site Specific Sound.* Errant Bodies Press.

LaBelle, Brandon. 2010. *Acoustic Territories: Sound Culture and Everyday Life.* Continuum.

Lefebvre, Henri. 1991. *The Production of Space.* Translated by Donald Nicholson-Smith. Basil Blackwell.

Lemmon, Eric. 2019. "Telematic Music vs. Networked Music: Distinguishing Between Cybernetic Aspirations and Technological Music-Making." *Journal of Network Music and Arts* 1 (1). https://commons.library.stonybrook.edu/jonma/vol1/iss1/2.

Merleau-Ponty, Maurice. 1962. *Phenomenology of Perception.* Translated by Colin Smith. Routledge & Kegan Paul.

Notes from the performance of *A Certain Geography* at "Forms of Ventriloquism." IMT Gallery, London, July 2014.

Oliveros, Pauline. 2009. "From Telephone to High Speed Internet: A Brief History of My Tele-Musical Performances." *Leonardo Music Journal* 19: 2–5.

Pinker, Steven. 1994. *The Language Instinct.* Penguin Books.

Schaeffer, Pierre. 2017. "Acousmatics." In *Audio Culture: Readings in Modern Music* (revised edition), edited by Christoph Cox and Daniel Warner. Bloomsbury.

Simpson, Paul. 2009. "Falling on Deaf Ears: A Postphenomenology of Sonorous Presence." *Environment and Planning* 41 (11): 2556–75.

Truax, Barry. 2002. "Genres and Techniques of Soundscape Composition as Developed at Simon Fraser University." *Organised Sound* 7 (1): 5–14.

15

Walking (and Paddling and Flying) with Heritage

Archaeologists' Daughters' Mobilities Across Northern Landscapes Captured on Disposable Cameras

Oula Seitsonen and Sanna Seitsonen
With Contributions by Sohvi Seitsonen, Elsa Seitsonen, and Elvi Seitsonen[1]

Archaeology is a unique science in that the entire family—from toddler to teenager to grandparent—can participate in and contribute to fieldwork and research (see Casimiro et al. under review; Seitsonen et al. 2021). As archaeology can study all traces left by human activities, both past and present, it can be approached rather intuitively. During field research, tasks can be assigned in ways that are suitable for all ages, allowing different communities to engage in research in various ways. Including one's family in fieldwork has clear advantages for the well-being of both the archaeologists and their relatives, generating positive experiences of being and doing things together. It also familiarizes children with their parents' profession—and, for many archaeologists, their deep passion (Casimiro et al. under review).

Both parents in our family are archaeologists, and before the children were born, we regularly worked together in Finland, Norway, and Russia. However, at present, only Oula is working full time in archaeology. Sanna participates in fieldwork about once a year to maintain a connection to the field. Our children, Sohvi, Elsa, and Elvi, have grown up around—and grown used to—our annual field trips and eagerly look forward to them each year. At the time of the writing, they are seventeen, fifteen, and nine years old, respectively, and have been involved in fieldwork for most of their lives. During their first field trips,

Sohvi was two years old, Elsa was six months, and Elvi was ten months old. This chapter was written jointly by Oula and Sanna, with important contributions and input from all our daughters.

Our joint family fieldwork has covered a wide geographic range of archaeological sites and contexts, primarily between sixty-eight and seventy-one degrees north (figure 15.1). These range from ancient Sámi habitation sites to nineteenth-century Lapland Gold Rush areas, and from Second World War ruins to twentieth-century Finnish-Canadian immigrant homesteads. In addition to work in our home country of Finland, we have conducted joint fieldwork in Sweden, Norway, Estonia, and Canada. Much of our research has taken place in Sápmi, the homeland of the Sámi people, the European Union's only Indigenous group. This has allowed us to introduce the girls to local Northern concepts of indigeneity, Indigenous livelihoods, and cosmologies through ongoing interactions with our Sámi friends and colleagues. It has also given us opportunities to discuss various (traditional) Northern lifeways, worldviews, and perceptions of landscape, which often differ from the dominant (Southern) perspectives (for example, Magga 2022).

Over the years, we have developed a tradition of bringing disposable cameras along on our fieldwork (Seitsonen 2021; figure 15.1). For children today, a manual camera with actual film and a limited number of frames represents a novel, exciting, and antiquated technology. It requires a different and more measured approach to seeing, photographing, and documenting activities and landscapes compared to using a mobile phone or digital camera. For one, the physical process of taking photographs with an analog camera differs from digital photography and is more time consuming. One must manually advance the film, raise the camera to the eye to peer through the viewfinder, and carefully consider focus and lighting (which are often automated in digital cameras) before finally pressing the shutter button to take a picture. This protracted process stands in stark contrast to the casual motion of snapping numerous quick shots with a mobile phone at arm's length, which many people do almost without thinking about it.

Capturing mobility and movement in two-dimensional still images has been a challenge since the pioneering days of photography (for example, Prodger 2003). Simple disposable cameras face some of the same challenges that early photographers encountered, due to the limited ability to adjust image-capturing settings and quality, aside from using the built-in flash. This becomes even more interesting when considered at a theoretical level. At the core of a photograph's

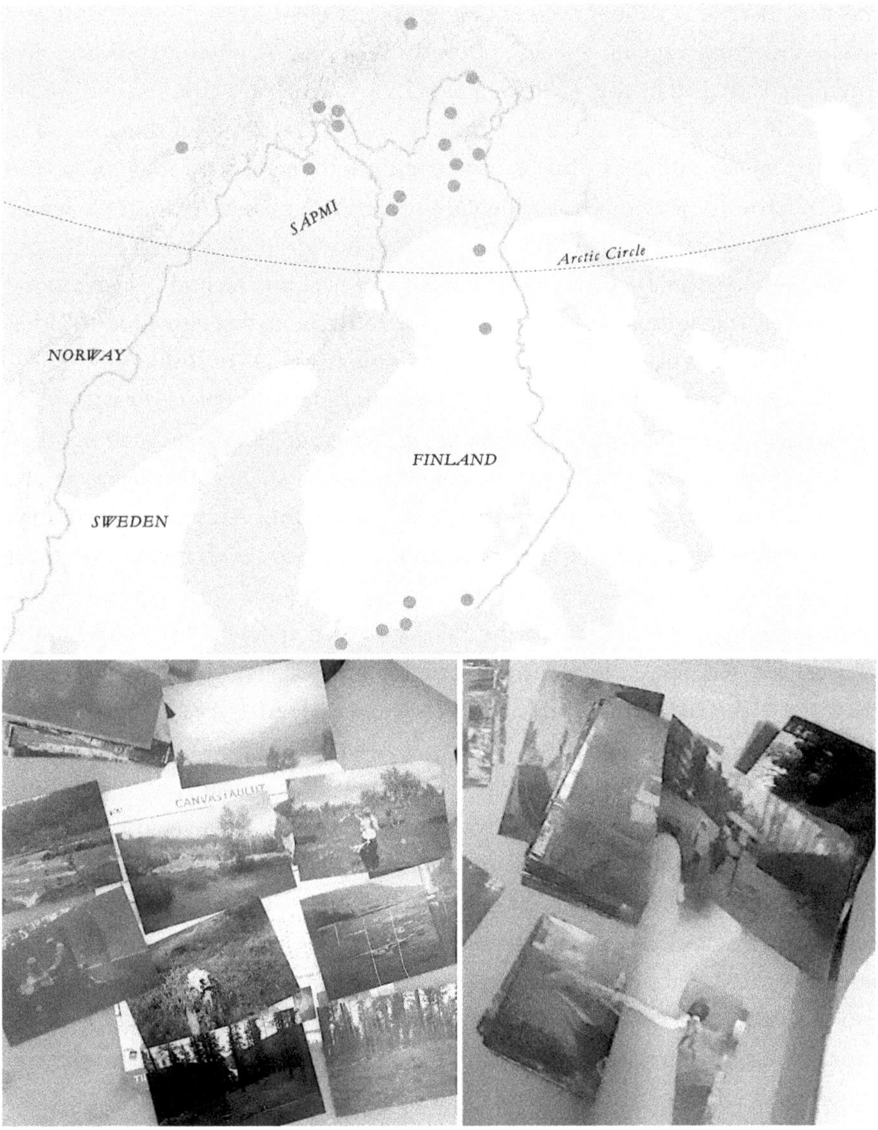

Figure 15.1 *Top*. Distribution of our family's archaeological fieldwork locations from 2009 to 2024 (marked with red dots; the Sápmi region is shown in pink), primarily above the Arctic Circle. In addition to Finland, Sweden, and Norway (shown on the map), Sohvi, Elsa, and Elvi have also participated in fieldwork in Estonia and Thunder Bay, Canada. (Background map: Esri/ArcGIS Pro) *Bottom*. Looking together at the girls' photographs spread out on the sofa and floor, recalling where they were taken, what happened at that moment, and what was going on outside the picture frame. Source: Photo by Oula Seitsonen.

material essence lies the movement of a photographed scene through time and space, involving varying spatiotemporal transitions. A photograph, first and foremost, is shaped by the movement of photons, which are captured and frozen at a specific moment in space and time. These images are then transported in material form—either as film or developed photographs—through space and time by various conscious or accidental processes (Seppänen 2014, 169). Space-time shifts are thus ontologically inherent to photographic representation, and particularly personally meaningful images can evoke perceptions of temporal and spatial transitions, creating a sense of movement through time to other temporalities and places—a kind of mental time travel (2014, 169).

It is important to note that photography and photographs carry with them a historical burden of the objectifying (male) gaze, voyeurism, and colonial records (Kind 2013), particularly in colonized areas such as the Sápmi region (Lehtola 2018). Today this typically manifests as the touristic gaze, representing what outsiders perceive as an empty natural wilderness, even though for locals it has always been—and remains—their ancestral cultural landscape and home (see Seitsonen et al. 2021). Northerners approach their cultural and natural heritage, as well as their home landscapes, from a very different viewpoint than the "Western gaze" is able to perceive. The faint traces of past human activities often remain invisible to outsiders not trained to see and interpret them. The traditional Sámi worldview emphasizes reciprocal relationships and cohabitation in the landscape with both human and nonhuman actors, and their ethical principles of land use aim to leave minimal marks of past activities on the landscape (for example, Tervaniemi and Magga 2019). On our expeditions, we have consciously endeavored to use our photographs to engage with an interconnected and corporeally experienced world—an approach inspired by Sylvia Kind's (2013) work, which involves children reciprocally in photographic experiences. This process has the potential to make children's inner worlds and experiences visible, offering insights into them (Kind 2013, 428).

In this chapter, we present a collaborative photographic essay about Sohvi, Elsa, and Elvi Seitsonen's field photography in Arctic landscapes. All the photographs in the chapter were selected by the girls themselves. While choosing the images for the essay, we discussed how their photographs reflect, for instance, the mobilities across Northern heritage landscapes and their own perceptions of archaeology, our joint family fieldwork, and the world more broadly. These mobilities were achieved through various methods, from walking to canoeing to flying in a helicopter. In what follows, we consider these as they appear in Sohvi's, Elsa's, and Elvi's imagery. This essay is autoethnographic and highly personal in nature. Autoethnographic methodologies are inherently defined by high degrees of subjective personal engagement and emotional attachment. When examined through analytical reflection and introspection, they can yield deeper understandings of various subjects (for example, Ellis et al. 2011; Ikäheimo 2020).

Even with all the modern technological advances, such as swift flying transitions offered by air transport and aerial or satellite remote sensing surveys, in the end, all archaeological fieldwork comes down to walking the soil and confirming observations firsthand on the ground. The fictional archaeologist and desert explorer Count Almásy, based on a real-life historical figure, aptly commented on this in the movie *The English Patient*, which is based on Michael Ondaatje's 1992 novel of the same name: "You can't explore from the air, Madox. If you could explore from the air, life would be very simple" (Minghella 1996). When conducting an archaeological exploration, one must engage closely and corporeally with the land, traversing the landscape and encountering the world at one's feet—often with one's eyes fixed on the ground to spot even the tiniest hints of past human activities, such as burned bone fragments or waste flakes from Stone Age tool production (for example, Bell 2020; Bell and Leary 2020).

Walking is an exemplary mode of mobility for making sense of our world, experiencing it personally, and connecting and interacting with landscapes (for example, Graves-Brown and Schofield 2019; Ingold 2004, 2010). Walking in and through our lived environments and lifeworlds creates and transforms these spaces and places through our corporeal interaction as we move along (for example, Ingold 2004, 2010; Seamon 2018). Landscapes are dynamic entities, continuously shaping and shaped by human movement and interaction. Cultural meanings and a sense of place are actively constructed, reconstructed, and inscribed on them over time, along people's "lines of becoming" (Ingold 2011), which form connected temporal meshworks woven by human activities (for example, Ingold 1993, 2016; Klenk 2018). Past mobilities are a common theme in archaeological studies, often explored by modeling and reconstructing the accessibility of various archaeological landscapes, features, and structures (for example, Aldred 2014, 2021; Leary 2014; Morton et al. 2012, 2013; Seitsonen and Viljanmaa 2021; Seitsonen et al. 2014). However, mobility as a field method has been less frequently discussed by archaeologists (see Graves-Brown and Schofield 2019), despite its high potential for more in-depth analyses and intimate engagement with field sites (see Herva et al. forthcoming 2025).

Working together with our daughters while looking through their photographs was very enjoyable, much like the fieldwork captured in them. Spreading the pictures across the floor and browsing through them was engaging, enlightening, and fun. We showed the photographs to each other, chatting about them, pointing out details, and remembering where they were taken, what happened before and after the shots, and what the wider circumstances around them were that made them interesting enough to be captured. All this created a shared experience for us. Reflecting on it now, it feels as though this process opened new perspectives on the expeditions and our memories for everyone in the family.

All quotes beneath the photographs come from Sohvi, Elsa, and Elvi, as heard over the years while walking in the field or looking through images from past fieldwork at home (the girls have requested that the quotes not be personalized with respect to who said what). The first author considered their quotes, comments, and shared experience from a more theoretical perspective and attempted to explain these reflections to the girls in an accessible way, allowing them to correct or refine them. All family members have agreed on the final version of this chapter. The reader is encouraged to read the image captions carefully and compare them with the photographs to reflect on the girls' comments about the selected scenes.

For adult observers, the photographs taken by Sohvi, Elsa, and Elvi, along with the discussions about them, open up—both literally and figuratively—new perspectives on places, things, and themes, including movement, mobilities, and accessibility. Children, being physically shorter in stature, quite literally see the world from a different angle. Thus, their pedestrian gaze (Graves-Brown and Schofield 2019), or their optic array (Gibson 1979), differs from that of adults. As they walk through a landscape, children—depending on their age and size—observe the relationships, shapes, and superposition of various elements from different angles and at different speeds compared to both adults and older and younger children. Figuratively, children also tend to observe the world in a distinct way, encountering it with greater open-mindedness and perception. These perspectival differences are evident in the photographs taken by Sohvi, Elsa, and Elvi. The arrangement of these photographs and the various elements within them reflect their differing perceptions of the world. This is particularly true of Elvi's photographs, as she, being the youngest, has not yet been as influenced or constrained by the pictorial conventions of society, typified by the almost ever-present and constant image flows we experience today (Seitsonen et al. 2021).

Many of the girls' images convey a powerful sense of elevation and perceptual depth, created by the impressive and immersive rugged, mountainous landscapes, as seen in figures 15.2 through 15.4 (see also Seitsonen et al. 2021). These images, along with the very experience of slowly moving through the vast landscapes, highlight the smallness of human presence in the face of "untamed" Northern nature. This serves as a good reminder, even for adults, who may have forgotten about it in their human-centered, anthropocentric worlds. As one of the girls commented, "Nature is big and silent in the mountains."

Relying on traditional, slow-paced forms of mobility during fieldwork—largely walking, but also paddling along the rivers in a rubber boat, canoe, or kayak—further strengthens the sense of smallness in the vast Northern wilderness. When working in the field with children, one must consider their individual abilities to perform daily tasks and plan accordingly. For instance, one should not count on traversing thirty kilometers of terrain in a day while

Figure 15.2 "Nature is big and silent in the mountains. And cold when the wind is strong and it feels almost like winter, then you need all the warm clothes and good rain gear." Elsa and Elvi wander across the tundra in the pouring rain in 2020. Our joint fieldwork has introduced the girls to places and conditions often otherwise unknown to children of their age, and has also prepared them to cope with such environments. The first author has sought to familiarize the girls with places that are important to him for professional, personal, and phenomenological reasons. Source: Photo by Sohvi Seitsonen.

Figure 15.3 "There was a mobile phone connection on the top, so I could message my friends and show the scenery." The value of being outside of mobile phone coverage is something the first author has consciously emphasized to the girls during family field trips. Not atypical for their generation, the girls almost always download video and audio files to their phones before going off-grid. However, there is still much more leisure time in the wilderness — for games and reading in tents and cabins, for example. In this image from July 2023, Sohvi negotiates her way across a slippery patch of perennial snow and ice above the clouds on Finland's "rooftop" — the renowned Háldi Mountain in northernmost Sápmi. Walking across unfamiliar and challenging terrain keeps one acutely aware of bodily movements and requires constant concentration to avoid slipping or injuring oneself. This makes one evermore conscious of both the act of moving and one's connection with the land. Source: Photo by Elsa Seitsonen.

carrying a thirty-kilogram backpack, as Oula is sometimes prone to doing. The distances to be covered, unanticipated weather changes, and sometimes challenging terrain conditions are much more of a concern when one is out with children than when working with trained adults. The responsibility for the children's well-being weighs heavily on the parents' shoulders in wilderness settings, particularly when there is limited or no access to immediate medical care—even with a satellite phone as a backup one can't count on evacuation if something goes wrong; one has to be prepared to deal with different situations and scenarios at least in the first place on one's own. Moving across rugged

Figure 15.4 "We lost the GPS device as we climbed to the rocky, open fjell top. Then Daddy had to navigate in the old-fashioned way with a map and compass to get us back to the cabin. I think I'd get lost in the forests and fjells without guidance, with only the reindeer tracks to follow." In 2021 Elvi and Sanna make their way down the steep slope leading to the Ivalojoki Kultala Crown Station in this 19-century Gold Rush landscape in Lapland. Learning essential skills for traditional mobilities is something we have focused on while in the field together—everything from finding one's way across different types of terrain safely to understanding the basics of using paper maps so as to make sense of elevation contours and compass navigation. Source: Photo by Elsa Seitsonen.

terrains requires learning specific skills for navigating various settings and conditions, as well as understanding the potential hazards and risks involved, such as loose boulders, thick undergrowth, sudden gorges, slippery cliffs, and so on. Teaching and learning these movement skills, along with other useful and vital wilderness skill sets—like setting up camp in various conditions or cooking with a camp stove—have also been an important part of our family fieldwork, from the parents' perspective.

Movement—or tropes of mobility—appears in various ways in most of the girls' photographs. For instance, one can see them traversing the tundra into the faraway distance (figure 15.2), slowly and carefully negotiating their way across barren wet rocks and slippery snowbanks (figure 15.3), climbing the steep slopes of remote river valleys among massive ancient pines (figure 15.4), or walking on and touching the summertime snow patches in wonder (figure 15.5). The often stormy, cloudy skies also seem to convey a sense of being alive and in constant motion (figures 15.2 and 15.3). Even in the more serene scenes, you

Figure 15.5 "This was just before Elvi fell into the spring while filling a water bag." Photographs of feet and legs illustrate movement in their own specific way, even when the feet are still, as they are in this image. Every element in the photograph from 2021 conveys a different kind of movement at a different scale: the stream flows swiftly from the spring where we got our drinking water, running beneath the rickety, makeshift bridge; the wet vegetation swaying gently in the wind; and Elsa's rubber boots, briefly immobile, reflect in their glistening surface the clouds racing across the skies above the Avviljohka (Ivalojoki) River. Source: Photo by Sohvi Seitsonen.

can sense the world on the move—for example, the rapidly flowing stream below the wobbly bridge where Elsa has stopped (figure 15.4), with the swiftly moving clouds reflected in her rubber boots; or the graceful movements of reindeer grazing among the softly swaying mountain birches (figure 15.7); or the gentle ripples of the current behind our raft on the Avviljohka River (figure 15.8). Indeed, the only element in the pictures selected by the girls that feels completely motionless and lifeless is the landed helicopter in figure 15.9—highlighting a tension between the various natural movements in the other pictures and this mechanized, silent metal construct.

From discussions with them, it is clear that the girls feel their photographs and the interconnected memories from our expeditions represent pleasant memories. They enjoy displaying the photos on the walls and sometimes pull them out to browse through (figure 15.10). These images seem to have subconscious metonymical properties for them, serving as reminders of good

Figure 15.6 "Seeing and touching snow in the summertime is one of the most exciting things; the summer is so short up there. We were also sliding down the jassa (the North Sámi word for perennial snow patches)." Summer snow patches are a curiosity that the first author fondly remembers from his own childhood, when he would slide down them with his sister. He felt joy at seeing his own children share this experience (Seitsonen et al. 2021). "There is no vegetation this high up, rocks are loose and bare, and there is no protection at all from the constant wind." In the 2023 photograph, Sanna, Oula, Sohvi, and Elvi survey along a recently melted snow patch, with the light gray, lichen-free rocks revealing the limit of the area previously covered by perennial snow and ice (Seitsonen et al. in press 2025). Photo by Elsa Seitsonen.

times and exciting experiences in nature in various parts of the world. As they described it, "the pictures bring back and maintain memories that are good to have." It has been fascinating to observe how the shared fond memories of new experiences or stunning mountain scenery have, in their minds, overshadowed any unpleasant aspects of the fieldwork, such as the sometimes taxing daily treks, slippery trails, chilling winds, or freezing rain and fog that often beset us during these moments.

From Oula's perspective, the most fascinating aspect of writing this essay has been observing the shared process of engaging corporeally with the paper prints—handling them, examining them closely, pointing out their details, and reminiscing together about the memories, emotions, and insights they awaken. This demonstrates the dynamic, controversial, and fluid nature of photographs as

Figure 15.7 "It's nice when reindeer come to visit our excavations; they seem curious about what we're doing. Reindeer are pretty, and their hooves make a funny sound when they walk. They also sound a little like piggies when they snort." Multispecies interactions are a significant part of fieldwork in the North, especially with reindeer but also with other species, such as birds of prey, wolverines, fish, and other animals we have encountered, as well as with the plant life that often hides archaeological traces beneath it, and with inanimate beings, such as the mythical "Stallo" giants of Sámi folklore (Seitsonen et al. 2021). Past human and animal activities have shaped the flora, and these differences in vegetation patterns can indicate the presence of archaeological sites, as seen here near Lake Gilbbesjávri (Kilpisjärvi) in 2022 (see Seitsonen and Égüez 2022). Source: Photo by Elvi Seitsonen.

physical objects (Seppänen 2014, 96). Photographs are imbued with contrasting elements, with memories that are simultaneously present and absent, and capture fragments of a past that can remind us of a larger experience. While these fragments evoke nostalgia and awaken memories, they also constantly remind us that the original moments are irretrievably gone (Seppänen 2014, 96–97).

Physical interaction with paper prints appears to serve as a sensory trigger, actively and vividly bringing back memories and forming new connections between the images and the remembered actions and landscapes they depict.

Figure 15.8 "Paddling the Avviljohka River was fun but slow. This was where we could use the motor when the water was deeper. We saw eagles, reindeer, and other animals along the way." During our joint fieldwork in 2021, we followed the route taken by gold prospectors during the late-1800s Gold Rush in Lapland, paddling down the river. Re-enacting the relaxed pace of past mobilities, paddling opened up different perspectives on movement compared to motorized transport—even though, in some places with deeper water, we were able to use the outboard motor, as shown in the image, which also allowed Sohvi to take a break from paddling and snap the image. On the left, Elvi is seen taking a picture with her own disposable camera. The blurriness of the image captures the experience of swiftly bustling down the rapidly flowing river, as well as the limitations of disposable camera technology when in motion. *From left to right*: Elvi and Sanna, our expert guide Peeter Veltson, and our colleague Suzie Thomas. Source: Photo by Sohvi Seitsonen.

In our experience, staring at immaterial digital imagery on a flickering computer or mobile device screen does not elicit an analogous engagement (Seitsonen et al. 2021; Seppänen 2014, 192). This can be interpreted through the phenomenological lens proposed, for instance, by philosopher Maurice Merleau-Ponty (1962, 161), who suggested that when we reach out or move toward an object with our body, our "general instrument of . . . comprehension" (1962, 235), we are responding to that object's invitation to interact. This idea can be applied to paper photographs, as the image we touch invites us to engage not

Figure 15.9 "Flying in the helicopter to the mountains and paddling down the Avviljohka River were the most exciting experiences. Here, Daddy and the pilot are unloading our gear at Háldi." Our joint fieldwork has given us some experiences that would otherwise have been beyond our reach, such as flying by helicopter to survey the remote wilderness areas. This mode of transport is more cost-effective than doing the four-day hike to the area before even reaching the research site. At right, Oula and the pilot unload the helicopter in 2022, before the pilot departs; he will return to pick up the family after their land surveying work (weather permitting). Weather is unpredictable in the high mountains, which must always be considered when making fieldwork plans (see Herva et al. 2025), especially when going out into the field with children. Source: Photo by Elvi Seitsonen.

only with what is visible within the frame but also with the unseen elements we can imagine beyond the picture borders, prompted by the personal experiences conjured by our intertwined memories.

All in all, it was wonderful to share the immersive experience of flipping through the two-dimensional images, witnessing how these thin paper prints could evoke memories beyond the metonymical scenes within each picture frame, and how each scene could grow to represent the memory of a past, three-dimensional, lived world, experienced bodily (Seppänen 2014, 52). While looking closely at the familiar landscapes and events in the pictures, one could begin to feel as though one were being pulled into the image—or, in Merleau-Ponty's (1962) terms, answering their call for interaction. During this process, it felt as though one could in fact metaphorically step together into the photographed scenery, and in one's mind's eye, look around to see the captured moment through shared memories. Staring thoughtfully at some of the powerful images, one could even begin to sense the physical presence of the porous pasts

Figure 15.10 "I like to put pictures from excavations on my walls; they remind me of the trips and places. I change them sometimes—right now, there are photographs from Avviljohka above my desk and from Megonjávri on the wall next to my bed, so I can see them." Our whole family enjoys looking at our photographs after "fielding" as reminders of the times we had, even if in the rainy, snowy, and windy field itself it is not always so pleasant. The girls have expressed that they value tangible, material reminders from our field trips—of physically moving across and engaging with the landscapes—rather than a bunch of immaterial digital files lingering on a phone, as is too often the case nowadays. Source: Photo by Oula Seitsonen.

documented by them. Slowly, one might begin to hear the distant, familiar hum of the swaying dwarf birches and the soft melody of the crystal-clear mountain brooks in the otherwise silent solitude of the wilderness; suddenly, one might feel below one's feet the rugged rocky terrain once walked, sense the faint whiff of tundra vegetation, and turn to face the chilling cold wind and driving rain—all in a metaphorical walk into one's own past.

Acknowledgments

We would like to express our thanks to all our colleagues who have joined (and tolerated) our family in the field, especially Anu Herva, Vesa-Pekka Herva, Carol Kajorinne, Shawn Morton, Kerkko Nordqvist, Roger Norum, Meaghan Peuramäki-Brown, and Suzie Thomas. This paper is part of the University of Oulu projects "Eerie Arctic Aviation" (Research Council of Finland, decision number 363043, 2024–2028) and "An Archaeological Perspective on Inequality in Welfare Society" (Kone Foundation, 2022–2027).

Note

1 Our chapter is a family effort by Sanna, Sohvi, Elsa, and Elvi Seitsonen based on our joint experiences and expeditions, as we all enjoy walking and exploring the world together.

References

Aldred, Oscar. 2014. "An Archaeology of Movement: A Methodological Study." PhD diss., University of Iceland.

Aldred, Oscar. 2021. *The Archaeology of Movement*. Routledge.

Bell, Martin. 2020. *Making One's Way in the World: The Footprints and Trackways of Prehistoric People*. Oxbow.

Bell, Martin, and Jim Leary. 2020. "Pathways to Past Ways: A Positive Approach to Routeways and Mobility." *Antiquity* 94 (377): 1349–59. https://doi.org/10.15184/aqy.2020.133.

Casimiro, Tânia M., Joel Santos, Afonso Leão, Gustav Wollentz, Hanna Steyne, Oula Seitsonen, Sarah Bockmeyer, and Susana Pacheco. Under review. "Archaeological Osmosis, or How Is Archaeology Seen by Those Who Share Our Lives?" *Open Archaeology*.

Ellis, C., T. E. Adams, and A. P. Bochner. 2011. "Autoethnography: An Overview." *Forum Qualitative Sozialforschung/Forum: Qualitative Social Research* 12 (1): 10. http://nbn- resolving.de/urn:nbn:de:0114-fqs1101108

Gibson, J.J. 1979. *The Ecological Approach to Visual Perception*. Boston: Houghton Mifflin.

Graves-Brown, Paul, and John Schofield. 2019. "Encountering Landscape: Travel as Method." *Landscapes* 20 (1): 61–84. https://doi.org/10.1080/14662035.2020.1800272.

Herva, Vesa-Pekka, Aki Hakonen, Roger Norum, Oula Seitsonen, and Markus Fjellström. 2025. *Weirding Landscapes: Arctic Glacier Extinction and the Otherworldly Monsters of the Anthropocene*. Palgrave Macmillan.

Ikäheimo, J. 2020. "The Rescue Excavation and Reburial of Late Pet Animals as Explorative Archaeological Autoethnography." In *Entangled Beliefs and Rituals. Religion in Finland and Sápmi from Stone Age to Contemporary Times. MASF 8*, edited by T. Äikäs and S. Lipkin, 236–252. Helsinki: Archaeological Society of Finland.

Ingold, Tim. 1993. "The Temporality of the Landscape." *World Archaeology* 25 (2): 152–74. http://www.jstor.org/stable/124811.

Ingold, Tim. 2004. "Culture on the Ground: The World Perceived through the Feet." *Journal of Material Culture* 9: 315–40. https://doi.org/10.1177/1359183504046896.

Ingold, Tim. 2010. "Footprints through the Weather-World: Walking, Breathing, Knowing." *Journal of the Royal Anthropological Institute* 16: 121–39. https://doi.org/10.1111/j.1467-9655.2010.01613.x.

Ingold, Tim. 2011. *Being Alive: Essays on Movement, Knowledge and Description*. Routledge.

Ingold, Tim. 2016. *The Life of Lines*. Routledge. http://refhub.elsevier.com/S1462-9011(18)30327-7/sbref0140.

Kind, Sylvia. 2013. "Lively Entanglements: The Doings, Movements and Enactments of Photography." *Global Studies of Childhood* 3 (4): 427–41. https://doi.org/10.2304/gsch.2013.3.4.427.

Klenk, Nicole. 2018. "From Network to Meshwork: Becoming Attuned to Difference in Transdisciplinary Environmental Research Encounters." *Environmental Science and Policy* 89: 315–21. https://doi.org/10.1016/j.envsci.2018.08.007.

Leary, Jim. 2014. *Past Mobilities: Archaeological Approaches to Movement and Mobility*. Ashgate.

Lehtola, Veli-Pekka. 2018. "Our Histories in the Photographs of the Others: Sámi Approaches to Visual Materials in Archives." *Journal of Aesthetics and Culture* 10 (4): 1510647. https://doi.org/10.1080/20004214.2018.1510647.

Magga, Päivi. 2022. "Defining the Sámi Cultural Environment." In *The Sámi World*, edited by S. Valkonen, Á. Aikio, S. Alakorva, and S.-M. Magga. Routledge.

Merleau-Ponty, Maurice. 1962. *Phenomenology of Perception*. Routledge.

Minghella, Anthony. 1996. *The English Patient*. Miramax/Tiger Moth Productions.

Morton, Shawn G., Meaghan M. Peuramaki-Brown, Peter J. Dawson, and Jeffrey D. Seibert. 2012. "Civic and Household Community Relationships at Teotihuacan, Mexico: A Space Syntax Approach." *Cambridge Archaeological Journal* 22 (3): 387–400. https://doi.org/10.1017/S0959774312000467.

Morton, Shawn G., Meaghan M. Peuramaki-Brown, Peter J. Dawson, and Jeffrey D. Seibert. 2013. "Peopling the Past: Interpreting Models for Pedestrian Movement in Ancient Civic-Ceremonial Centres." In *Mapping Spatial Relations, Their Perceptions and Dynamics: Lecture Notes in Geoinformation and Cartography*, edited by S. Rau and E. Schönherr. Springer. https://doi.org/10.1007/978-3-319-00993-3_3.

Prodger, Phillip. 2003. *Time Stands Still: Muybridge and the Instantaneous Photography Movement*. Oxford University Press.

Seamon, David. 2018. *Life Takes Place: Phenomenology, Lifeworlds, and Place Making*. Routledge.

Seitsonen, Oula. 2021. "Rain, Reindeer, Digging and Tundra: Children's Visual Perception of an Archaeological Expedition to Northernmost Sápmi (Finnish Lapland)." *Time and Mind* 14 (3): 431–57. https://doi.org/10.1080/1751696X.2021.1951563.

Seitsonen, Oula, and Natalia Égüez. 2022. "Here Be Reindeer: Geoarchaeological Approaches to the Transspecies Lifeworlds of the Sámi Reindeer Herder Camps on the Tundra." In *ISKOS 24. Materiality and Objects: Multi-Disciplinary Approaches to Archaeological Material and Contexts*, edited by P. Halinen and J. Taivainen. Suomen Muinaismuistoyhdistys.

Seitsonen, Oula, and Sami Viljanmaa. 2021. "Transnational Landscapes of Sámi Reindeer: Domestication and Herding in Northernmost Europe 700–1800 A.D." *Journal of Field Archaeology* 46 (3): 172–91.

Seitsonen, Oula, Jean-Luc Houle, and Lee Broderick. 2014. "GIS Approaches to Past Mobility and Accessibility: An Example from the Bronze Age Khanuy Valley, Mongolia." In *Past Mobilities: Archaeological Approaches to Movement and Mobility*, edited by Jim Leary. Ashgate.

Seitsonen, Oula, Markus Fjellstöm, Aki Hakonen, Vesa-Pekka Herva, Roger Norum, Sanna Seitsonen, Sohvi Seitsonen, Elsa Seitsonen, and Elvi Seitsonen. 2025. "Archaeology of Finland's Only 'Permanent Glacier': The Past, Present, and Future of the Ritničohkka Snowfield, Sápmi." *Journal of Glacial Archaeology*.

Seitsonen, Sohvi, Elsa Seitsonen, and Elvi Seitsonen 2021. "Megonjávri Expedition Photo Gallery." Supplementary material to Seitsonen (2021). Figshare. https://doi.org/10.6084/m9.figshare.14975160.v2.

Seppänen, Janne. 2014. *Levoton valokuva*. Vastapaino.

Tervaniemi, S., and Magga, P. 2019. Belonging to Sápmi: Sámi conceptions of home and home region. In *Knowing from the indigenous North: Sámi approaches to History, Politics and Belonging*, edited by T. H. Eriksen, S. Valkonen and J. Valkonen, 75–90. Routledge.

16

Walking Interrupted

Camilla Brudin Borg

During a planned research trip to investigate stories of pilgrimage on the Camino del Norte in Spain, an injury forced me to stop and remain in a place that pilgrims typically pass quickly by. From my stranded position, I watched groups of pilgrims with their eyes firmly fixed on their destination: the city of St. James, Santiago de Compostela. "Buen camino!" I shouted encouragingly, but I could not join the march myself. My planned week long walk was over, cut short by my injury; but, as it turned out, my research journey was not. Instead this interruption led to an unexpectedly close encounter with the Costa Vasca Geopark on the Basque coast, and a new sense of participation in broader perspectives—where the planet, evolutionary time, prehistoric relatives, and mythical stories appeared side by side with my own walk in the present.[1] In this text, I will recount the discoveries and experiences that resulted from the interruption of my long-distance walk, influenced by techniques of zooming out, and explore how an interrupted pilgrim might reconnect with the different layers of time and existence embedded in the landscape.

Modern Pilgrimages in Northern Spain

Most research on modern pilgrims emphasizes human existential and social reasons as the driving forces behind their journeys. Many pilgrims find themselves at a personal crossroads, experiencing a crisis that calls for respite, a change of scenery, and spiritual guidance. They walk for many reasons. Often they are motivated by internal human concerns or the longing for a renewed connection with the divine; sometimes they are driven by the desire to be part of the community (*communitas*) and to experience a sense of belonging to a "camino family," a temporary but strong group constellation that often forms along the pilgrimage route (Gemzöe 2016, 125). Many pilgrims also imagine themselves following in the footsteps of their medieval predecessors, giving them the feeling of belonging to a human community that spans centuries. However, reasons such as the reevaluation of one's connection with Mother (or Father) Earth are rarely mentioned as motivations for a pilgrimage, even though the experience of nature is often cited as a powerful side effect of the walk (Frey 1998, 38–41). Pilgrimage, then, seems to be primarily an existential exercise and an anthropocentrically motivated project that takes place in a socially defined human present, with perhaps a medieval historical horizon as a backdrop.

The almost automatic anthropocentric framing of pilgrimage is an important part of both the practice itself and the scientific paradigm through which it has been studied. However, the inherent anthropocentrism in both the experience and the academic study of pilgrimage can be problematized—particularly by adopting methodologies that use movement, proximity, and corporeality (Kinnunen et al. 2024). One can also turn to feminist new materialist theories, material ecocriticism, and Indigenous research, which are based on ontological assumptions that emphasize humans' cocreation and interaction with the more-than-human world (Iovino and Oppermann 2014; Kimmerer 2016). Other fields, such as postcolonialism, poststructuralist cultural criticism, and geocriticism, underscore the importance of the spatial dimension in meaning-making and thus the need for a broader analytical focus.

Along these lines, in *Sense of Place and Sense of Planet* (2008), Ursula K. Heise calls for a shift in perspective, a theoretical "zooming out" to a global and planetary level, in order to visualize how the local and global are interwoven. Similarly, geology's attempt to coin the concept of the Anthropocene ("the age of humans") has prompted humanities researchers to zoom out in deep time and to a planetary level (Crutzen and Stoermer 2000). As a result, we are increasingly moving beyond a one-sided, human-centered view to one that also embraces multispecies, socially pluralistic perspectives grounded in cocreated thinking and action—also in relation to the planet. Therefore it is intriguing to consider whether the practice of pilgrimage can be connected to nonanthropocentric perspectives and to the understanding of our interconnectedness with Earth's other species and ecosystems in deep time.

Although pilgrimage has traditionally focused on the salvation of the human soul, both the Camino Frances and the Camino del Norte provide ample opportunities for pilgrims to zoom out and engage with deep historical and

Figure 16.1 Flysch formations at Zumaia Beach, Camino del Norte. Source: Photo by Camilla Brudin Borg.

planetary perspectives. Both routes begin in the mountain range between France and Spain, winding through the prehistoric landscapes of the Pyrenees. As they walk, pilgrims constantly encounter magnificent and interesting sites where history—spanning vast geological and evolutionary contexts across epochs and eons—unfolds tactilely directly beneath their feet. Along the Camino Frances, pilgrims pass Atapuerca, where the remains of Europe's earliest human ancestors have been found. On the northern route, pilgrims pass the famous Altamira and Ekain caves, known for their advanced prehistoric art, as well as the Costa Vasca Geopark, renowned for what geologists call its flysch formations (from the German *fließen*, meaning "to flow"). These flysch layers are rhythmically interwoven sedimentary stones, five to twenty centimeters thick, that record the various historical periods. The rocks appear as differently colored layers of fine-grained sediments—shales and mudstones—arranged in perfect order, now weathered into jagged formations resembling neat rows of stone teeth. The slow, gradual erosion of the sea against the rocks, with some layers being harder than others, has sculpted these distinct, dragon-scale-like formations. The Basque flysch formations represent various eras and phases of Earth's development, with the remains of each period—reflecting its unique climate—arranged like bundles of differently colored fabric or books placed neatly on a shelf, waiting to be read.

Geologically, the area was shaped when the Iberian Peninsula rotated and collided with the European continent between eighty and twenty million years ago. This collision, which compressed the landmasses, created the Pyrenean Mountain range. The seabed's horizontal layers of fossilized sediments, stacked in kilometer-thick strata that represent millions of years of geological history, were gradually twisted into a vertical position—which is how we still encounter them today, especially along the stretch between the villages of Zumaia and Mutriko.

The entire area has great geological and geomorphological value and has been protected by UNESCO since 2015. Managed by the Costa Vasca Geopark, this spectacular natural heritage has been entrusted to the Geogarapen organization by the towns of Zumaia, Deba, and Mutriku for safekeeping (Monge-Ganuzas 2021, 106). The clear traces of the planet's evolution offer walkers a tactile experience of time; the strange landscape can not only be "read" but also felt. In contrast to the vast time embodied by the stone, the pilgrim appears as a tiny creature, stumbling across a vast planet that was born millions and millions of years ago and floats in infinite space. Is it possible to understand our human role within this immense context? Can the individual pilgrim relate to—or at least get a brief, breathless glimpse of—their own place (or perhaps insignificance) in this infinite vastness? In this article, I will walk, read, and attempt to engage with this remarkable place, with its material and mythical traces of deep narrative time.

Storytelling on the Camino del Norte

It is worth examining how writing pilgrims experience the landscape and environments while walking the Camino del Norte. What significance do they attach to space, place, and time? If we look at pilgrims' reports focused on the first part of the Camino del Norte, we find that there are not many printed publications describing the pilgrimage along this route. As I was preparing to walk the Camino del Norte myself, I brought along Jean-Christophe Rufin's *The Santiago Pilgrimage: Walking the Immortal Way* (2016), in which the French

Academy member describes his arduous adventures on the northern route, and Finnish-Swedish journalist Stefan Lundberg's travelogue *På Spaniens tak: En vandring utan slut* (On the Roof of Spain: A Journey Without End, 2017). Both begin their walk in the twin towns of Hendaye/Irun, located on either side of the French-Spanish border. This is the starting point for many pilgrims who set out along the northern route.

In both Rufin's and Lundberg's work, it is striking that direct descriptions of the rock formations along the coastline are largely absent. Lundberg introduces the arrival on the Camino with a few sketchy brushstrokes that stylishly convey height, views, and airiness: "Every ten minutes I exhale with relief, lean on the walking sticks, take a sip, and look out over the rolling hillside that loses itself in the Atlantic. . . . Far below, a yacht draws a white arc across the blue Bay of Biscay. . . . The view, the wind, and the smells are intoxicating" (Lundberg 2017, 12, trans. Brudin Borg).[2] This passage effectively conveys the physical experience of sweaty, hard hiking, the scents, and the windy breezes, thus creating a clear sense of place. Lundberg also mentions the wonders along the Costa Vasca, but only briefly: "I pass Zumaia with its dramatic beaches and hundreds of surfers and walk along beautiful roads towards the town of Deba" (2017, 28). In the same breath, he notes: "An open landscape with valleys and mountains is better for the flight of thought than a sea horizon. The mountains give perspective and depth. I think" (2017, 28).[3] The landscape Lundberg portrays with impressionistic ease will become a suitable, though anonymous, backdrop for his walking reflections, retrospective musings, and historical contextualizations. However, these interwoven thoughts are often set in places other than the Camino del Norte.

Rufin, on the other hand, is deeply preoccupied with his emotions as he starts out along the Camino del Norte. He soon learns to look for the yellow Camino signposts that call out "this way, this way": "In a vast landscape, full of details, of foregrounds and backgrounds, the pilgrim's eye, like radar, instantly detects the marker stone, the arrow, the sign that points to Santiago. . . . But on this first morning, I was still a beginner. A novice reader, unaccustomed to deciphering the Saint-James script" (Rufin 2016, 34f). Rufin quickly becomes a "reader" of the road signs. The arrows point him in the right direction, guiding him through landscapes, cities, and varied terrain. Similarly, Lundberg, in his own account, reflects on the function of the arrows, describing the experience of following the yellow signs as "almost like a game . . . I will learn to see them in the most peculiar places" (2017, 11).[4]

Figure 16.2 Signs showing the way to Santiago de Compostela. Source: Photo by Camilla Brudin Borg.

As he approaches the village of Zumaia and the Costa Vasca Geopark, Lundberg gives us a bird's-eye view. In his book, two photos show the coastal landscape, accompanied by the caption, "Spain's wild north coast offers graphically beautiful landscapes" (Lundberg 2017).[5] Here, once again, the aesthetic qualities of the mountains and the expansive, airy perspective of the view are emphasized. The description is, of course, accurate; there are many opportunities to admire long, unbroken panoramic vistas along the Camino del Norte. In Lundberg's observation, there is a direct connection between the free flight of thought and the design of the landscape itself, but the prehistoric dimensions and a close geological reading of Zumaia with its dramatic beaches are missing. Lundberg dwells mostly in his own mind.

However, he has done his homework, offering a historical background on Spain, the Basque Country, and the Camino, with details about their histories and specific sites, much like in a journalist's travel account. At times, his reflections turn personal, but the magnificent landscape and its strange formations do not really seem to capture Lundberg's interest. The fact that he literally walks across such a massive chapter of Earth's history for an entire day goes unnoticed. Instead the pilgrim Lundberg moves primarily within the anthropocentric, human-centered condition and in his own private thoughts.

In contrast, when Rufin boldly tackles the first stages of the Camino del Norte, ranting and raving somewhat, he soon realizes that this is a physical undertaking to be taken seriously. He complains that he has lost his identity. To him, pilgrimage life seems focused entirely on pushing forward and satisfying simple, basic needs. On the first day, he feels as though he is losing his sense of direction, his vanity, and the dignity that his titles and social status provide. The project seems, to him, to border on madness, but he concludes that he still understands the value of the ascetic enterprise of walking eight hundred kilometers over rough terrain—and that he probably needs to lose everything in order to understand what is truly essential.

The geological surroundings are more present in Rufin's story than in Lundberg's, and we are drawn into the green Basque landscape with its spectacular sea views and rock formations. Camping in a bay, probably in Algorri or possibly near Deba, Rufin writes:

> It was on this stretch on the Way that I found the most unusual places to camp. I spent one night in a narrow cove between two cliffs, a place where layers of rock, weathered by costal erosion, descend like the teeth of a giant comb into the

> tousled hair of the waves. Parallel lines of pink and grey rock spread out from the shoreline towards the horizon. When the tide is out, you can walk along this magical causeway paved with rocks, with little lines of water rippling between them. I was blessed with a glorious sunset. The last rays of the sun spread their light all the way from the horizon to where I stood, following the straight tracks marked out by the rocks just above the surface of the sea. (Rufin 2016, 51)

There is no trace of geological reflection here; instead Rufin primarily expresses an aesthetic, wondering appreciation of the place's beauty. The famous flysch formations of the coastal strip, however, are obviously the backdrop. He also zooms in on the strange seabed plateaus exposed at low tide. Rufin reaches for a metaphor to capture the meeting of mountain and water, likening the rock to "the teeth of a giant comb" that is descended "into the tousled hair of the waves." Yet this analogy does not go beyond aesthetic contemplation. In contrast, Rufin experiences the powerful agency of the place through the harsh treatment of the weather, spending a shaky night in a roaring storm in his little tent.

In Rufin's and Lundberg's retellings of their pilgrimages, the external landscape largely recedes into the background, appearing only in brief glimpses. The narrative is primarily concerned with the inner lives of the walkers. The outside world is depicted as pale, airy, aesthetic—or prosaically reduced to the search for yellow way markers. In *Landscape and Power*, W. J. T. Mitchell argues that "landscape" is really a fantasy, an image shaped by various practices, laws, and regulations (Mitchell 2002), similar to how Henri Lefebvre describes "the conceived space" (Lefebvre [1974] 1991, 38–40). In this context, the pilgrims do not really seem to *see* the strange formations of Costa Vasca, or, like Rufin, quickly pass by them. Instead they engage in practices of aesthetic seeing or overtly focused seeing. When they look out from their inner landscapes, they are moved by panoramic views—but it is as if they see their surroundings without truly seeing. We are left to ask why.

Distance and Pace

Anyone setting out on a pilgrimage to Santiago de Compostela will be walking a long distance. The minimum requirement to obtain a *Compostela* certificate is to walk at least one hundred kilometers on foot, but many pilgrims walk up to eight hundred kilometers. Modern pilgrims increasingly express a desire to receive a certificate that indicates the exact distance they have walked. As a result, the pilgrimage office in Santiago de Compostela now issues a "Certificate of Distance" for those who wish to have it. This practice suggests that distance has taken on special significance and value, with pilgrims eager to prove that they have covered a specific, considerable distance. The long pilgrimage, then, stands apart from shorter, day long meditative walks, primarily because of its distant goal, which defines the rhythm of each day of the journey.

The pilgrim follows a well-marked path, guided by yellow markers. For many, a purposeful, forward-driven movement—always in search of the next yellow marker—seems to determine their speed and pace. Amit and Salazar (2020, 2) emphasize that movement is always a matter of pace, duration, intensity, and timing. The walking project itself is framed by an external set of ideologies, practices, and norms that influence how the walker moves through and relates to the landscape and surrounding world (Salazar and Glick Schiller 2014). We should also keep in mind that, in order for the long-distance pilgrim to arrive within a set time—often dictated by the duration of their vacation—a certain distance must be traveled every day in a disciplined way. The pace must be maintained, and the rhythm adjusted to cover the daily distance. Pilgrims may form social connections along the way, but newly formed groups must find a common pace—a prerequisite for keeping the group together. When a pilgrim agrees to be accompanied, choosing or changing the pace, and deciding when to stop and take breaks, is largely out of their control, and they may not be able to engage deeply with the villages and places along the route. In this way, pace and distance play a crucial role in how the pilgrim engages with their environment, and the enforced tempo of the journey can shape both what they see and experience.

Not surprisingly, the medieval Latin word *peregrinus* (from Latin *peregrinor*, meaning "to travel abroad") denotes a stranger or vagabond, evoking images of exile. A pilgrim wanders from village to village but remains a stranger to the locals. Constantly on the move, with the destination always in mind, a pilgrim rarely stops for long.

In my own interrupted state, I began to question the impact of this alienating, goal-oriented way of moving. How does it affect the way we see, experience, and relate to our surroundings—whether to animals, villages, or the broader planetary context? What changes when the pilgrimage is interrupted, when the speed is altered, or when we stop altogether? For me, the goal of reaching the cathedral in Santiago was replaced by questions prompted by a slow, rather than urgent, pace; by wandering and curious seeing, rather than the goal-orientated search for yellow trail markers.

Interruption: Change of Method

Originally, the aim of my research project was to make participatory observations on the Camino del Norte by walking with other pilgrims, conducting walk-and-talk interviews, and identifying patterns and similarities with the narratives of Jacobean literature. When it became clear that I could no longer carry out such a walk, I needed to find a new method that better suited my changed situation, a method that could take advantage of proximity, slowness, and interrupted movement. "Method comes from the Greek *meta* and *hodos* and means to be on the way. And that, ultimately, is the key here, both in literature and in life, to learn to experience the dynamic movement of life, to be in the method, in a constant state of being" (Wingård 2024, 58).

Walking researchers Springgay and Truman (2018) use a relational, embodied, and ethically mindful walking method that allows the researcher to "think in motion." They draw on Isabelle Stengers's ideas about "the politics of slowness," where slow movement is "speculative" and seen as a "process of unlearning and unsettling what has come before" (15)—learning with the world, questioning the practice of learning about the world. This approach creates a relational and attentive walking method that builds on Juanita Sundberg's concept of "walking-with," offering an ethically and politically responsible way of walking alongside other epistemic worlds (Sundberg 2014, 35)—a method also explored in several chapters of this anthology. It draws from new materialism, affect theory, and Indigenous, critical, and environmental theory, which all describe intertwined relationships with more-than-humans. The method requires, above all, slowness and an openness to emergence, allowing for recognition of an ongoing intra-action with the material word. In *Ethics and Politics of Space for the Anthropocene*, Rantala and Valtonen (2020) further develop the method by, among other things, practicing walking with stones and creating a micro-level contextual analysis focused on care for the more-than-human. I now sought to adopt this care-focused, slow, and proximity-oriented method to engage with the place on this planet where I, as a pilgrim, was stranded and interrupted.

Reading in a Planetary Way

The ancient Greek philosopher Protagoras famously stated, "Of all things the measure is Man, of the things that are, that they are, and of the things that are not, that they are not," as discussed in Plato's *Theaetetus* (Pt. 152a2–4; Diels and Kranz 1952). Protagoras' *Homo mensura* argument points to the inevitable relativism of human thought and measurement. But are we doomed to be forever confined to a short sighted, anthropocentric perspective? In *The Great Derangement* (2016), Amitav Ghosh argues that we must zoom out in time and space to place the individual narrative within a larger context, in order to fully understand how we are connected at a planetary level. After all, it is only within these broader dimensions and scales that long-term trajectories such as climate change, slow violence, and global injustices can be truly understood.

In his book *Stone: An Ecology of the Inhuman* (2015), Jeffrey Jerome Cohen introduces the concept of "tactile archive" (77)—time that can be touched, that bears traces of its passage. Stones, he argues, are narrative matter or "polychronic agents" (91) that are able to bridge enormous time gaps when encountering humans. This way of reading can activate a vision that spans thousands, or even millions, of years.

On the coast of the Basque Country, with the sea to the right and towering mountains in the background, pilgrims are invited to see and feel the planet's evolutionary history in deep time through the very stones and strange formations around them. This becomes particularly evident outside the village of Zumaia, with its steep cliffs of flysch formations, palimpsestic layers of petrified time. Hands and feet can gently trace over the five historical moments when all life on Earth almost perished. Here humans become contemporaneous with the dying invertebrates and the macroscopic waves of time.

To transform out of my narrow pilgrimage world, I began to learn about evolution, to listen to the archaic language of ancient life; I exchanged my near-sightedness for a zoomed-out perspective. I stepped into a boat and followed the rhythmic, dramatic coastline with my body, experiencing the clashes of sliding continents. A bus took me to Saturraran Beach, where the first black flysch can be found, created over one hundred million years ago. In Mutriku's Nautilus Geological Centre, I encountered the giant ammonites—fossils of strange and beautiful creatures that once inhabited the Earth and served as nests for early squid species.

On the impressive Sakoneta Beach lies what is known as "the Andutz fault," visible as a sharp boundary between black and white flysch, marking the meeting point of the Lower Cretaceous (also called the Old Cretaceous) and the Upper Cretaceous (the Younger Cretaceous). A little farther east lies the cove known as the Algorri, where Jean-Christophe Rufin likely camped during a stormy night, and where it is possible to read in Nature's book about the dramatic event that occurred sixty-six million years ago, when a ten-kilometer-wide meteorite struck the Chicxulub crater in the Yucatán Gulf in the Caribbean Sea, causing earthquakes, tsunamis, and large wildfires. In addition, the story in the rock tells of an Earth that was shrouded in an impenetrable particle cloud, resulting in the extinction of 76 percent of Earth's life, including the dinosaurs. This mass extinction, the fifth in Earth's history, is visible here as a millimeter-thin black line in the cove's rock formations. I run my hand over the black stripe, becoming simultaneous with the event. It took over ten million years for the Earth's ecosystem to recover. What was once the lush Cretaceous period transitioned into the new geological era of the Paleocene.

Now the flysch stripes begin to turn red, and I walk on, reflecting. Today we are in the middle of the next mass extinction—this time caused by the destructive, colonial, and insatiable lifestyle developed in the Global North, which is now spreading like a dark poison across the world. The epoch we live in may be called the Anthropocene—the age of humans—but the eon of the planet's evolution in which we find ourselves is called the Phanerozoic. Phanerozoic means "visible life"; it began 541 million years ago, when multicellular animals first evolved—our ancestors, and by extension, us.

Indeed, Zumaia, Deba, and Mutriko are all recommended stops in pilgrimage guides, but pilgrims tend to pass through these villages quickly on their way to Santiago. The magnificent landscape and the historical traces of deep time shouldn't leave any passerby indifferent; however, as we saw in Lundberg's and Rufin's stories, the imprints of more than one hundred million years of the planet's history often fail to make it into autobiographical, self-centered narratives. The expanding space that pilgrims enter in these regions requires a different way of seeing—one that does not place the individual, the self, or even the social at the center. Yet, paradoxically, pilgrims often carry with them a desire to create a new narrative about themselves. This opens up the possibility for a different kind of walking narrative that transcends the private and existential, and instead reads "us" alongside the macro- and micro-scales of time and space that unfold with the rhythm of the waves of stone between Zumaia and Mutriko. What is needed is close attention to the surface underfoot, but also a zooming out in the imagination to the walking space traversed on planet Earth.

When the Apollo spacecraft sent back its first distant images of our planetary home in the 1970s, the pictures captured what looked like a fragile "blue marble" floating in space. The perspective was shifted, and the point of view and scales became "astronomical," prompting a renegotiation of our role as humans (Heise 2008). The images of the whole planet embraced by space are familiar to everyone now, but along with (the argued) concept of the Anthropocene and a growing awareness of the finite nature of Earth's resources, we may now be more ready for a long-term, zoomed-out epistemology.

Entering Dragonstone

So I slowed down and took a short, zoomed-in and zoomed-out walk down the steps to Izurun Beach. The stairs run exactly along the line that separates the Paleocene from the Eocene, a boundary made visible by the dramatic change in the Earth's climate that occurred. Temperatures rose by five to eight degrees, a change thought to have taken place over about twenty thousand years. Geologists refer to this event as the Paleocene-Eocene Thermal Maximum (PETM), and the red rocks of the beach are a direct result of this ancient climatic change. In the solidified space, twisted into a vertical background behind the relaxed and laughing tourists, the crucial events of space-time played out—events that offer our closest reference to what might happen if global temperatures increase by just a few degrees today. I reflect on our own smallness. Yet we are so powerful. In such a short time, we have developed a lifestyle that rivals the forces that have shaped the planet for millions of years.

By crossing just a few hundred meters past the sunbathing tourists to the quieter western part of the beach, I move almost ten million years back in time. I decide to take a dip just outside one of the beach's mythical caves, formed by wave erosion. Jagged flysch formations are arranged in thin, fragile discs—razor sharp and archaic. The mountain's wounds open up like a monster's mouth. A strange feeling overcomes me. Have I been here before? I stare at the cave and am suddenly transported to another world in my imagination. My gaze sweeps out over the sea and back across the beach. The dragon princess Daenerys Targaryen is returning to her birthplace. I find myself at the very entrance to Dragonstone Castle in the HBO series *Game of Thrones* (Benioff and Weiss 2017), based on the fantasy epic *A Song of Ice and Fire* by George R. R. Martin (Martin 1996). Zumaia's dragon-shaped stony beach, with cliffs that themselves resemble dragons rising from the sand, has been used to evoke the fantasy setting of Dragonstone, superimposing this spectacular prehistoric site onto the realm of fantasy stories and fairytale adventures.

The Darkness in the Depths of the Cave

Fairy tales about mountains, stones, and caves often tells of shadowy creatures, of petrified trolls who cannot stand the light of day, of evil that dwells in the depths beneath the stone. Here, on the beach, it is not only modern fantasy that has incorporated the stone into the walker's storyworld; Zumaia's local oral tradition has its own stories:

> Legend had it that many years ago, a desperate Zumaiana who had not been able to have children, could not help but cry out in the wind, "I want a daughter even if at the age of 20 the devil is going to take her away." After nine months, the woman gave birth to a beautiful baby girl. When she was about to turn 20, the panicked woman locked her up, standing guard day and night. But at the very moment of her twentieth birthday, dark noises came from the room where the young woman was locked up, alerting the woman. When the woman entered the room in terror, she saw a dark figure running away, forcibly dragging her daughter. The whole town rushed to chase them, surprised when they saw the devil and the girl go into the Cave of Izurun. Confident of their capture, they rushed into the small cave, but there was no longer any trace of the girl who had disappeared forever. (Senditur n.d.)

Being "taken by the mountain," like the girl in the story, perhaps expresses a universal fear of being swallowed by destructive forces or by the unknown. In *Game of Thrones*, Princess Daenerys and the hero Jon Snow discover engraved images of "the white walkers" within the cave. In the fantasy world of *Game of Thrones*, the white walkers represent the all-consuming forces of darkness against which everyone must unite if the world is to survive.

Fairytales and fantasy are allegorical stories that reflect the subconscious of their cultures. They portray the struggle between good and evil, incorporating actual threats to our world through alienation and simplification. In this way, J. R. R. Tolkien's *The Lord of the Rings* (1954/1995) thematizes the madness that swept Europe during the two world wars of the twentieth century. Deep caves and mountains play an important role in depicting the underlying darkness in *The Lord of the Rings*. The mountains are rich with metals and precious stones, and these, along with the great craftsmanship of its dwellers, help them flourish. Yet the story also explores greed, the unchecked industrial exploitation of raw materials, and people's inability to be content. Greed summons the sleeping Balrog—a demon from the depths, a force that should have remained undisturbed. The dwarves are punished for their avarice: their culture is destroyed, and the mountain, hollowed out like a rat's nest, is left abandoned, decayed, and overtaken by the servants of evil.

Tolkien's allegorical critique of modernity, with its relentless industrialization, has become increasingly relevant today. The story's themes of greed for gold and precious stones are mirrored by today's extraction of the rare earth metals essential for the green transition (or green colonialism). As in Tolkien's mythical tale, we are likely approaching a tipping point in this new, unbridled greed—a limit to what the planet can tolerate.

Archaic layers of meaning in fairytales take on renewed relevance when placed alongside the stones' stories of prehistoric mass extinctions and climate change. Deep time, present time, and fantasy time converge on a beach in the Basque Country. The climatic collapse of past geological ages slides like flysch over the fantasy world's struggle against the forces that threaten all life. The dragons and the beach's geological formations become one. My pilgrimage has been transformed, and I now see through time's layers like a palimpsest. Climate collapse from millions of years ago and five previous mass extinctions—each wiping out (almost) all life on Earth—are visible here as thin lines in the ground beneath my feet. We are now in the midst of our own creation of the sixth mass extinction, with a very real risk that we, the human animal, will be among those erased from existence.

Notes

1 Professor Alasdair Skelton, professor of geochemistry and petrology at Stockholm University, has reviewed this article and confirmed that the geological facts are consistent with current research. The Royal Society of Arts and Sciences in Gothenburg had generously supported the research for this article.

2 "Var tionde minut pustar jag ut, lutar mig mot gångstavarna, tar en klunk och tittar ut över den böljande sluttningen som förlorar sig i Atlanten. /. . . / Långt därnere ritar en lustjakt en vit båge över den blå Biscayabukten. /. . . / Utsikten, vinden och dofterna är berusande." [All translations from Swedish to English are my own unless other is stated.]

3 "Jag passerar Zumaia med sina dramatiska stränder och hundratals surfare och går längst vackra vägar mot staden Deba." I samma andetag noterar han att: "Ett öppet landskap med dalar och berg är bättre för tankens flykt än en havshorisont. Bergen ger perspektiv och djup. Tycker jag."

4 / . . . / "som en lek att följa gula pilar" / . . . / "Jag kommer att lära mig se dem på de mest märkliga ställen."

5 "Spaniens vilda nordkust bjuder på grafiskt vackra landskap."

References

Amit, Vered, and Noel B. Salazar. 2020. *Pacing Mobilities: Timing, Intensity, Tempo and Durance of Human Movements*. Berghahn.

Benioff, David, and D. B. Weiss. 2017. *Game of Thrones*. Season 7. HBO.

Cohen, Jeffrey Jerome. 2015. *Stone: An Ecology of the Inhuman*. University of Minnesota Press.

Crutzen, Paul J., and Eugene Stoermer. 2000. "Have We Entered the 'Anthropocene'?" *Global Change Newsletter* 4 (5): 7–18.

Diels, Hermann, and Walther Kranz. 1952. *Die Fragmente der Vorsokratiker*. Weidmann. Frey, Nancy Louise. 1998. *Pilgrim Stories: On and Off the Road to Santiago. Journeys Along an Ancient Way in Modern Spain*. University of California Press.

Frey, Nancy Louise. 1998. *Pilgrim stories: On and off the road to Santiago*, University of California press.

Gemzöe, Lena. 2016. "I skrivande pilgrimers fotspår." In *Fiktion och verklighet*, edited by Eva Johansson and Tomas Lunden. Makadam.

Ghosh, Amitav. 2016. *The Great Derangement*. The University of Chicago Press.

Heise, Ursula K. 2008. *Sense of Place and Sense of Planet. The Environmental Imagination of the Global*. Oxford University Press.

Iovino, Serenella, and Serpil Oppermann. 2014. *Material Ecocriticism*. Indiana University Press.

Kimmerer, Robin Wall. 2016. *Gathering Moss. A Natural and Cultural History of Mosses*. Oregon State University Press.

Kinnunen, Veera, Outi Rantala, and Emily Höckert. 2024. *Researching with Proximity: Relational Methodologies for the Anthropocene*. Springer Nature.

Lefebvre, Henri. (1974) 1991. *The Production of Space*. Translated by Donald Nicholson-Smith. Basil Blackwell.

Lundberg, Stefan. 2017. *På Spaniens tak: En vandring utan slut*. Helsingfors.

Mitchell, W. J. T. 2002. "Preface to the Second Edition." In *Landscape and Power*, edited by W. J. T. Mitchell. University of Chicago Press.

Monge-Ganuzas, M. 2020. "Geoconservation in the Autonomous Region of the Basque Country (Spain)." *Geoheritage* 13: 106.

Rufin, Jean-Christophe. 2016. *The Santiago Pilgrimage: Walking the Immortal Way*. Translated by Martina Dervis. MacLehose Press.

Salazar, Noel B., and Nina Glick Schiller, eds. 2014. *Regimes of Mobility: Imaginaries and Relationalities of Power*. Routledge.
Senditur. n.d. "Georoute of Algorri. Legend." Accessed February 15, 2025. https://www.senditur.com/en/route/georoute-of-algorri/.
Springgay, Stephanie, and Sarah E. Truman. 2018. *Walking Methodologies in a More-than-Human World: WalkingLab*. Routledge.
Sundberg, Juanita. 2014. "Decolonizing Posthumanist Geographies." *Cultural Geographies* 21 (1): 33–47.
Tolkien, J.R.R. *Lord of the Rings*. 1954/1995. Harper Collins Publisher.
Valtonen, Anu, and Outi Rantala. 2020. "Introduction: Reimagining Ways of Talking about the Anthropocene." In *Ethics and Politics of Space for the Anthropocene*, edited by Anu Valtonen, Outi Rantala, and Paolo Farah. Edward Elgar Publishing.
Wingård, Rikard. 2024. "Holistic Method as an Ecocritical Quest." In *Contemporary Ecocritical Methods*, edited by Camilla Brudin Borg, Rikard Wingård, and Jørgen Bruhn. Lexington Books.

17

Walking for Treasures

Scavenger Apps, Attention Management, and the Creation of Heritage Topographies

Ane Ohrvik

According to the World Economic Forum, there are now more smartphones than people in the world (Richter 2023). While not evenly distributed globally, mobile subscriptions for smartphones are rapidly approaching nine billion. Smartphones assist us in various forms of digital communication, entertainment, information retrieval, and physical exercise, and they help us organize and live our lives economically, socially, culturally, geographically, and temporally. Online banking, document storage, calendars, photos, map services, encyclopedias, and other resources—organized through individual digital applications—make virtually all conceivable information available to us today, provided we have sufficient battery life and internet access. In many respects, smartphones carry our lives.

Among the wide range of digital applications available, location-based recreational activities designed for smartphones have gained significant popularity. Commonly referred to as scavenger hunts, these apps effectively combine digital gaming elements with indoor or outdoor physical activities. These activities generally offer digitally available tasks through app technology, paired with physical movement—mostly walking—in specific locations determined by the chosen tasks. Global positioning system (GPS) technology is integrated as a locative aid in these activities. Playful adventures like treasure hunts, nature or city trails, detective games, and special themed trails offer almost endless variety. Furthermore, the effective way these apps combine experiential learning and knowledge dissemination has also made them very popular for educational purposes in schools and museums (see, for example, Gray 2007; Baalsrud Hauge et al. 2022; Richardson and Ritchie 2024).

Today numerous apps are available for free download to private mobile devices and smartphones, often with additional paid options for corporate and educational use, premium memberships, or in-app purchases for extra in-game items that enhance gameplay. The latter business model is exemplified by *Pokémon Go*, an augmented reality mobile game designed for users to locate, capture, train, and battle virtual Pokémon that appear in the player's real-world location. By 2020, four years after its release, the game had surpassed USD 3.6 billion in revenue and accumulated 576.7 million unique downloads worldwide (Chapple 2020).

My curiosity lies in understanding what such apps *do*. At a fundamental level, their function involves directing users, primarily on foot (although other modes of transportation may be used in more complex games), to specific places and goals. As such, these games represent a form of attention management and direction that participants consent to. Furthermore, the direction has an end goal, and it is the process leading to that goal, as well as what occurs at the destinations, that motivates this study. Far from being mere tools for entertainment and physical exercise, I argue that these apps have also become important instruments in modern heritagization processes by disseminating knowledge and raising public awareness of local history and cultural conditions.

In the examples I will discuss, my focus will be on how the location-based scavenger app Geocaching directs participants toward Norwegian holy wells—natural springs or wells in the landscape revered or considered special or sacred due to their supposed healing properties, religious significance, or association with popular practices and legends. Specifically, I will explore how the app frames both the information and the activities, including the roles of those who create a "cache" for a Saint Olav well and those who participate in finding it.

Scavenging

The release of GPS signals for civilian use by the US government in the mid-1980s, combined with technological advancements in mobile devices, initially facilitated the invention of digital scavenger hunts (Skinner et al. 2018, 98). Among them, Geocaching is, by its own account, known as the world's largest treasure hunt app. Since Dave Ulmer hid the first cache in Beavercreek, Oregon, USA, on May 3, 2000, Geocaching has grown to host more than three million active "geocaches" (physical items called caches), hidden in 191 countries across all continents, with millions of active "geocachers" (participants in the game) of all ages (Woodward 2024). In the Norwegian capital, Oslo, for instance, the Geocaching website reports that there are 6,702 geocaches registered.

Other popular hunt apps include Goosechase, Scavify, and Actionbound, all widely used for corporate and educational purposes. Scavify, in particular, is renowned for its corporate team-building activities and has a large user base, making it one of the most popular in its category. As new apps are designed at record speed to introduce variety and innovation, their popularity shifts (cf. Ohrvik 2020). Additionally, new national and local scavenger hunt apps are continuously being launched, often taking even greater account of local cultural and geographical factors in their design.

In the Norwegian context, Stolpejakten.no ("The Pole Hunt") exemplifies such a national scavenger hunt. Organized by county, city, and area, local sporting associations provide the digital platform with nature and city trails. Participants use their mobile phones (or other digital devices) to scan QR codes attached to physical poles placed in the landscape. Once a QR code is scanned, participants complete one pole on a specific trail consisting of multiple poles. When a QR code is registered, participants may also receive additional information about the location of the pole, often in the form of short historical details or stories connected to the place.

What most scavenger apps that focus on outdoor activities have in common is their emphasis on promoting physical activity, such as walking, and thus encouraging a healthy lifestyle (see, for example, Khamzina et al. 2020; Lee et al. 2021). Predefined routes with thematically or geographically related checkpoints constitute complete route units that contribute to specific movement patterns. The routes participants take may be short or long, and while there is no guarantee they will be completed, all encourage movement—they get people walking. More than that, they get people moving in particular ways. Precisely because the route checkpoints contain content that contributes to varied experiences, walking takes on an additional dimension of meaning as a form of "attentive walking."

Furthermore, because these apps facilitate and generate social communities both virtually and physically, their social and psychological benefits are also emphasized (Lee et al. 2021). Their digital design made them particularly relevant during the global COVID-19 pandemic, as they were already tailored to tackle physical restrictions through built-in digital technology. By late March 2020 (and possibly earlier), the first "The Great Stay at Home Scavenger Hunt" was designed to address local and national quarantine restrictions (Get Knit

Events 2024). The pandemic naturally encouraged developers and organizations to explore further digital scavenging interactions. In line with this, digital scavenging was also integrated into the new mainstream digital realities created by the pandemic, such as Zoom and Teams groups and communities, whether initiated in professional or private settings. Very soon after the start of the pandemic, institutions also began using scavenger hunts as outdoor "safe activities," where social distancing was ensured (Iadonisi 2020). The extent to which scavenger apps were used outdoors and in nature depended on local quarantine restrictions. In Norway, for example, people were never confined to their homes or prohibited from going out, which allowed scavenging in nature to continue as usual.

Many apps are designed democratically, allowing participants to create content themselves, though within the framework of rules and limitations set by the app creators. Participants can design trails, routes, hunts, or single "posts" (named differently depending on the app and game), which other users can access and provide feedback on within the app. This is also the case with Geocaching, where anyone can create a geocache, as long as it is identified with the coordinates of its location (the "waypoint"), published on the Geocaching community's website, and permission has been obtained to hide a cache in a public place. A geocacher searches for physically hidden treasures (caches) placed in public locations, whether in rural or urban landscapes, often in the form of small toys, key rings, coins, or other items the geocache owner (cache creator) has left behind in the cache container. The idea is that the items collected can be placed in other caches, allowing them to travel across large areas over time.

The dynamic and creative features of scavenging are crucial, as they encourage participants to become active contributors to both the recommended movements within the games and the content presented at each "post." In this way, everyone shares in the attention management.

App Technology as "Spatial Regimes"

A simple yet essential notion regarding technology is that every new piece of technology we apply to sense, experience, and disseminate information about the world will necessarily alter the relationship between users of the technology and the world they inhabit (Jørgensen 2017). These tools are not passive conduits but active mediators that transform spatial experiences and perceptual frameworks

(Jørgensen et al. 2022). Relating this perspective to locative app technology in scavenger hunts, questions emerge about how they direct the user's attention and contribute to movements and experiences. In his study of the locative gaming apps Ingress and Geocaching, Anthony Bak Buccitelli (2017) reflects on how these apps can be viewed as specific "spatial regimes":

> I want to suggest that we might usefully consider these games not just through the vehicle of individual user experience but also as spatial "regimes," value-encoded systems of power that play out in the individualized user's experience of space and place. In other words, we must consider the role that augmented reality games, as well as other kinds of locative media, play in setting up the everyday conditions under which users encounter and come to understand spaces and places, how these technologies allow users to shape those experiences, and how users respond to the conditions set forth in these platforms. (9–10)

When Norwegian holy wells, or wells attributed to the national saint Olav, appear as part of locative apps like Geocaching, this can be viewed as a form of place-making and value encoding. What might have been an undefined, seemingly unknown, or forgotten space among scavengers is reestablished as a place for new users through the attention directed at the wells when they are turned into caches for game participants.

The app technology contributes not only to attention management through physical practices but also to the systematic framing of the participants' valued content elements, which they come to associate with and experience at these places. The process by which the St. Olav wells are spatialized and imbued with meaning within the Geocaching app parallels the process characteristic of heritagization—the ways in which certain cultural elements are emphasized and valued as significant and representative of the past. Just as pasts are always under negotiation, so too is cultural heritage, as it is always strategically situated to reflect what people care about at any given time (Meskell 2015, 2; Johannsen and Ohrvik 2020). It is my contention that when the St. Olav wells in Norway become Geocaching caches, the value coding facilitated by the app paves the way for their heritagization.

Understanding the Geocaching app as "spatial regimes" that contribute to the value-encoding of Norwegian St. Olav wells may usefully be divided into two steps. The first involves attention management, where space is verbally and visually defined to aid the geocachers' movement through the landscape toward the end goal. The second concerns the elements that contribute to value-encoded place-making: the St. Olav wells. Studying this framing of spatial elements necessarily involves identifying the built-in technological framing of attention and content in the app, and how geocachers respond to it.

The basis for this study is eleven active caches, all with names pointing toward a water source, St. Olav, or both. These do not represent a complete list but are those that result from a simple search for "St. Olav well" on the Geocaching website. When a direct reference is made to a cache, it will be identified by its corresponding identification number on the Geocaching website.

Attention Management and Movement

Fundamental to many scavenger games is the simple premise that participants physically move between given points in an environment. For scavenger hunts to function at all, participants must accept this premise and ensure movement between the checkpoints. However, in many app-based games of this type, this premise is overshadowed by the actual game content: a riddle to be solved, information to be processed, questions to be answered, or built-in game elements in the app that need to be experienced. In such games, the act of walking may become merely a means to an end rather than a central element in itself. Consequently, neither the movement between points (or checkpoints) nor the physical environment one traverses—whether in a city or in nature—holds any prevalent significance or role. The most typical example of this is *Pokémon Go*, where virtual and fictional elements dictate where participants move in the physical environment.

In games such as Geocaching, this dynamic differs because it is the physical environment that guides the movement between points, aided by built-in elements. It is the landscape itself and specific characteristics and places in it that constitute the gaming elements and features. Unlike in *Pokémon Go*, these elements are constructed around physical conditions in the landscape, providing a specific type of attention management throughout the movement process. As a result, the physical conditions through which participants move become central to the game, and where and how they walk, as well as what they observe in the physical landscape, become important elements that are partly controlled within the game. Engaging in Geocaching thus involves a specific form of attentive walking.

This attentiveness can be compared to John Urry's concept of "the tourist gaze." Urry emphasizes that looking is a learned ability, where "gazing" is socioculturally framed and is a "performance that orders, shapes and classifies, rather than reflects the world" (Urry and Larsen 2011). As such, gazing is never "pure sight" but is conditioned by personal experiences and cultural content provided by others, which together enable specific ways of looking. While geocachers may not fit the traditional category of "tourist" when engaging in their outdoor gaming, they certainly rely on a conditioned "gaze," shaped both by their own situating and by the app. As we will see, this gaze contributes to framing the visual experience through a variety of elements. Additionally, this conditioned gaze, which influences where and how geocachers walk, can be compared to a specific "gait," as described by Susanne Österlund-Pötzsch. Drawing on Urry's work, Österlund-Pötzsch coined the term "tourist gait" to highlight how walkers, as tourists and explorers, move in specific ways, with a strong investment in their surroundings that is reflected in their movements (Österlund-Pötzsch 2018, 126–67). In the case of geocachers, I argue that the app's built-in elements frame these movements, creating a specific form of "gait"—attentive walking.

In many ways, the Geocaching application is essentially constructed as a site-identifying and place-making technology. The "geo" refers to the search for a specific geographic location, while the "cache" denotes the process of selecting and storing (caching) information about that location. The standard routine of creating a cache, hiding it, and searching for it is what defines Geocaching (Freese and Hargittai 2010; Skinner et al. 2018). Several built-in elements within the Geocaching app contribute to geographically directing participants' attentive movement through the landscape, three of which are the focus here: geographical coordinates, encrypted hints and riddles, and general descriptions of the route. Additionally, logged responses from visitors often include reactions to the route, which help guide other participants in their scavenging.

The essential focus on geographic location—the "geo"—through coordinates is the only form of direction that, when used exclusively, keeps participants' attention predominantly on the digital app rather than on the landscape they are moving through, as they work to match the coordinates. Two sets of coordinates are always provided. First, the geographical coordinates give the direction based on latitude and longitude. Second, participants may choose to be guided by the Universal Transverse Mercator (UTM) coordinate system, which divides the Earth into sixty zones, each six degrees of longitude wide. The zones define reference points for the UTM grid coordinates within each zone, with the grid coordinates given as distances in meters to the east and north. For example, to find the cache named "Olavskjelda" (St. Olav's well) at Ligrend in Telemark County using geographical coordinates, you would use N 59° 30.424 E 007° 58.344, as specified, while the UTM coordinates are 32V E 441833 N 6596965, as given by the cache owner (GC2TGE3).

While coordinates provide technical precision, they simultaneously abstract the rich, textural experience of the physical terrain, risking the reduction of mountains, forests, valleys, and urban environments to mere numerical intersections for geocachers. The embedded digital and intangible representation of the environment contrasts with traditional wayfinding methods such as landmarks, signposts, and local knowledge, privileging mathematical abstraction over embodied geographical understanding.

For coordinates to optimally direct geocachers' movements, they must be accurate. When they are not, participants' attention shifts from the abstract coordinates to a visual reading of the landscape. There are many instances where the cache owner has not been precise enough in registering the coordinates, or where geocachers have moved the physical cache container originally placed according to those coordinates. In such cases, visiting geocachers often inform others about the misleading coordinates and provide new ones, or replace misplaced caches so others can find them through the app.

The visual reading of the landscape is further assisted by the pictures posted by the cache owner on the cache page. These typically include a combination of close-up shots focusing on the St. Olav well and wider landscape photos or images of parts of the route to the well (see, for example, GC20QE9; GC2TGE3; GC37T4X; GC1D07T; GC1V020). As a directing tool, these photos serve as comparative checkpoints for geocachers, helping them ensure they are on the right path. Additionally, photos posted in the log by other geocachers, which appear in the photo gallery on the site, serve as further attention-directing resources. While these images provide guidance, they do not reveal the actual physical cache, which remains hidden.

Another built-in feature that links or orients scavengers toward the actual treasure hunt is the encrypted hints offered at many posts. The encryption key is available next to the hint and is the same for all hints. For frequent geocachers, this directs their focus toward employing a known methodology, while for less experienced geocachers, it presents a puzzling challenge (though, for those seeking a quicker solution, a simple click on the "decrypt" link bypasses the obstacle). What all hints share is that they are visual descriptions of directions or the route toward the cache, or descriptions of the place where the cache is stored. These range from simple clues like "benatr / orange" (GC1EC2B), "Fcehpr / Sprouce" (GC1D07T), or "Haqre fgbar / Under stone" (GC2EJYJ) to more detailed ones like "Zntargvp, envyvat, gbc yriry / magnetic, railing, top level" (GC6RHAG), or step-by-step guides toward the cache: "1. Unatvat / uratre v ra rvareohfx 2. Nccebnpu sebz hccre yriry / cå ulyyr yvgg arq sen gbccra 3. Nebhaq bar naq unys zrgre hc. / 1. Hanging in a Juniper bush 2. Approach from upper level on a shelf a bit down from the top 3. Around one and half metres up" (GCK73Z). Using the hints requires geocachers to pay close attention to the surroundings they are moving through. This game design element does not primarily aim to provide geocachers with a nature experience or immersion but instead instrumentalizes natural elements for in-game purposes. However, these hints sometimes direct attention toward the actual St. Olav well by placing a cache so close to it that the hints and the location merge into the same post—for instance, "nobir gur jryy / above the well" (GC2TGE3)—encouraging searches for the cache.

Many cache owners provide descriptions and details of how to navigate the landscape. These may include distances in meters from a central orientation point, such as a parking area (GCK73Z), the type of paths or routes one should take (GC20QE9), and potential physical obstacles in the terrain, along with advice on how to manage them (GC2EJYJ; GC6RHAG; GC1V020). These descriptions complement a built-in five-star terrain rating system, which indicates progressive levels of difficulty on the cache page. In addition, each cache owner registers so-called attributes associated with the cache, which are symbols communicating what to expect at the cache location. The attributes are divided into six categories, with each subsequent symbol offering guiding elements to help geocachers plan the trip and navigate the landscape. "Conditions" and "Danger" are two categories that offer guidance about the terrain. An experienced, attentive geocacher will use these symbols to assess the route's length, the potential for a difficult climb, whether the route and site are accessible in snowy conditions, or if there is a risk of encountering dangerous animals.

However, it is geocachers themselves who contribute the most in terms of providing fellow geocachers with directions by logging additional information about the route as they move through the landscape toward the cache. They add detailed descriptions of the landscape, as well as uploading personal photos of the well or nearby areas, further preparing other geocachers for the cultural experience that visiting the well may represent. As attention-managing tools, the landscape photos, the verbatim and symbolic descriptions by the cache owner, and the logged responses from participants all encourage geocachers to engage with the landscape, observe their surroundings, and interpret them according to specific guidelines. In this way, the digital format enhances the physical experience rather than competing with it, as is the case with some other scavenger games.

Scavenging as Heritagization

The attention management provided by the Geocaching app, which guides geocachers to St. Olav wells, generates new and often unexpected encounters and experiences with environments, narratives, and history, transforming the landscape into meaningful places. This transformation includes the valuation of St. Olav wells as significant sites, representative of the past as heritage—places worth learning about, remembering, sharing, and experiencing. In this context, the app functions as a dynamic container for place-lore. As expressions of place-lore, the geocache sites bring together locations, objects, legends, beliefs, memories, and practices in symbiotic relationships between communities and their environments (cf. Valk and Sävborg 2018, 9). In short, place-lore elements are essential in the value-encoded place-making associated with St. Olav wells (cf. Buccitelli 2017).

Whenever a cache is linked to a St. Olav well, the cache owner makes sure to explain why. Often the cache itself is named "St. Olav well," and the description that follows presents key elements that further identify and explain why the location was chosen:

> St Olav wells is a term used for a series of water sources in Norway. They are most often named in definite form, Olav's well, but one also finds the form St Olav's well. The wells were named after Olav the Holy in the years following Christianity's arrival in Norway, and are often located in connection with pilgrimage routes, old pilgrim paths. The water from these wells was considered holy and was believed to provide blessing and strength to those who drank from or touched it. Before this, it is assumed that many of the wells had connections to paganism, and rituals associated with Åsatru (Norse paganism). Several legends are therefore often attached to these wells. (GC2TGE3)

This textual excerpt from a geocache serves as a good illustration of how geocachers are typically introduced to a St. Olav well. The name of the well is explained through the physical properties of the water source, complemented by historical background that situates Olav Haraldsson within the context of Norway's Christianization and his sanctification following his martyrdom. Such historical narratives are often followed by presentations of legends linked to his name, recounting how, as both a historical and mythical figure, he was believed to have been present at the well's location at some point in history. These legends not only enhance the place's properties and significance but also serve to validate its function and historical significance. As legends, they invite their audience to consider their plausibility, thus creating their own unique realities in the landscape (cf. Oring 1986, 125; Ohrvik 2021, 112).

The same attention given to guiding geocachers through the landscape via the built-in game elements on the Geocaching app is also extended to guiding them historically and culturally when they arrive at the site:

> Legends says that Olav the Holy, during his Christian mission through Valdres in the year 1023, passed through the place while crossing to Land [a place name]. His troops were thirsty, and Olav thrust his sword into the ground. Water immediately sprang up freshly from beneath a stone, and the army quenched their thirst. The water in the Olav wells was supposed to have a healing effect (GC1D07T).

The narratives essentially convey how the wells are part of a sacred geography—one shaped by religious transitions, pilgrimage routes, and holy waters with healing properties—and how they hold ritual significance through the various cultural practices associated with them. As culturally embedded objects, they evolve into elements that, as Alexandra Walsham puts it, serve as "a mnemonic to their knowledge of previous eras, and a source of ideas about their social identity" (Walsham 2011, 6–7). By pointing to a memorable past while simultaneously remaining part of present-day practices, they embed layered mnemonic inscriptions that explain the who, what, and when of the well, thereby transforming it into a place of cultural significance and meaning.

When visitors digitally share their own experiences in the logs—describing how they ended up at the well, their journey to it, and their reflections upon arrival—they add additional layers of mnemonic inscriptions that contribute to establishing the well as a place. Many of the logged responses from visitors to these caches reveal that, without the associated geocache, they likely never would have visited the site: "I would probably never have found this place without the Geocache:D" (GCK73Z); "Finally I got to visit Olavskilden, which I have heard of but never got to visit before now; there had to be a cache for me to visit it" (GC2EJYJ); "Thank you to all the COs for doing the job of posting objects in places I probably never would have visited if it wasn't for your efforts" (GC4GE08); "Having lived almost my entire life just a few miles away, this is the first time I've been to the Olav's well" (GCK73Z). It is the Geocaching game that enables visits to and experiences of the place, and geocachers express clear gratitude for the inclusion of the St. Olav well on the list of geocaches. Geocachers also add historical information, share other legends or stories about St. Olav, and reflect on their own experiences at the site—activities that highlight how the technology allows users to shape experiences and respond to the conditions defined by the app platform (cf. Buccitelli 2017, 9–10).

A common behavior among geocachers is engaging in the ritualized practices connected to the site. References to the water's sacred and healing qualities are often described on the cache page, with accounts of how it has been part of prior practices and rituals. Many visitors report drinking the water, collecting it in bottles, or refraining from using it because it did not look healthy. Others throw coins in the well for good luck. All these testimonies point toward an acknowledgment of prior practices and an effort to engage and partake in them during their visit.

Geocachers don't refrain from documenting their visit in another common mode: the selfie. As a widely used way of creating and curating identity markers and sharing personal content socially, the selfie is also an important tool for validating experiences and sites. Pictures can serve as proof of presence, and judging by the many photos uploaded by geocachers at the cache sites, this documentation is important. With their own face in the foreground to indicate "I was here," many pictures capture the well in the background as the important object to be identified with. As such, the selfie can be seen as a display of the value-encoded place-making that has occurred.

Heritage Topographies

My initial claim was that scavenger apps like Geocaching have become important tools in modern processes of heritagization, as their built-in features facilitate these processes in very specific ways. When Norwegian holy wells associated with St. Olav appear on platforms like Geocaching, they undergo transformation. Previously obscure or forgotten locations are revived, evolving from undefined spaces into specific places, enriched by historical context, rituals, and compelling narratives. The app technology doesn't merely direct physical movement; it strategically curates experiential landscapes, imbuing sites with cultural significance. In doing so, it allows geocachers to experience landscapes in new and meaningful ways.

The app technology prepares and facilitates this process through "spatial regimes," which include a specific form of attentive walking toward the well, as well as the value-encoding that occurs at the site. Geocachers' "gaze" is conditioned by various built-in elements that guide them through the landscape. Some elements are numerical and technological, such as the coordinates, which shift geocachers' focus away from the landscape they traverse, while others require them to pay close attention to their surroundings—where they are and what they see. In this way, the app technology both abstracts the landscape and encourages careful observation of it. The way geocachers walk—their "gait"—is conditioned and curated by technology in similar ways. When they arrive at the site, the app functions as a dynamic place-lore container, facilitating the value-encoded place-making process on multiple levels and connecting the St. Olav well to history, narratives, rituals, and past and present experiences.

My initial question in this study was: What do these apps do? It is my contention that the Geocaching platform serves as a sophisticated tool for spatial recontextualization, where technological mediation enables historical sites to gain renewed relevance. Different app features and elements foster heritage-making dynamics, continually negotiating cultural meaning and strategically highlighting aspects that resonate with contemporary perspectives through active geocachers. By offering caches such as St. Olav wells, the Geocaching app maps and documents not only the physical features of a location but also the cultural, historical, and social meanings embedded in the landscape, thereby creating and curating specific heritage topographies.

References

Buccitelli, Anthony Bak 2017. "Locative Gaming, Folk Geographies, and the Experience of Cultural Heritage." *Cultural Analysis* 16.1: 7–29.

Chapple, Craig. 2020. “Pokémon GO Surpasses $3.6 Billion in Lifetime Revenue as It Celebrates Four-Year Launch Anniversary.” *Sensor Tower*, July. Accessed December 17, 2024. https://sensortower.com/blog/pokemon-go-revenue-year-four.

Freese, Jeremy, and Eszter Hargittai. 2010. “Cache Me If You Can.” *Contexts* 9 (4): 66–68.

Geocaching. Accessed December 16, 2024. https://www.geocaching.com/play.

Get Knit Events. 2024. “The Great Stay at Home Scavenger Hunt.” Accessed August 26, 2024. https://getknitevents.com/stay-at-home-scavenger-hunt/.

Gray, Howard Richard. 2007. “Geo-Caching: Place-Based Discovery of Virginia State Parks and Museums.” *The Journal of Museum Education* 32 (3): 287–93. https://doi.org/10.1080/10598650.2007.11510578.

Hauge, Jannicke Baalsruud, Esteban W. G. Clua, Jimmy Jaldemark, and Heinrich Söbke. 2022. “Educational Location-Based Interaction.” *Interaction Design & Architecture(s)* 52: 141–43. https://doi.org/10.55612/s-5002-052-001pfs.

Iadonisi, Matteo. 2020. “Fairmount Park’s Social-Distance Scavenger Hunt Unlocks Summer Fun During Pandemic.” *6 Action News*, August 12. Accessed December 16, 2024. https://6abc.com/timely-social-distance-open-space-nature/6367514/.

Johannsen, Dirk, and Ane Ohrvik. 2020. “How to Be a Pilgrim: Guidebooks on the Norwegian St Olav Ways and the Heritagization of Religion.” *Numen* 67 (5/6): 508–36.

Jørgensen, Finn Arne. 2017. “Walking with GPS: An Object Lesson.” In *Methodological Challenges in Nature-Culture and Environmental History Research*, edited by Jocelyn Thorpe, Stephanie Rutherford, and L. Anders Sandberg. Routledge.

Jørgensen, Finn Arne, Marie-Theres Fojuth, and Karen V. Lykke. 2022. “Innledning: Å finne veien—stedsansen som kulturell praksis.” *Tidsskrift for kulturforskning* 21 (2): 3–11.

Khamzina, Madina, Kaustubh V. Parab, Ruopeng An, Tiffany Bullard, and Diana S. Grigsby-Toussaint. 2020. “Impact of Pokémon Go on Physical Activity: A Systematic Review and Meta-Analysis.” *American Journal of Preventive Medicine* 58 (2): 270–82. https://doi.org/10.1016/j.amepre.2019.09.005.

Lee, Jung Eun, Nan Zeng, Yoonsin Oh, Daehyoung Lee, and Zan Gao. 2021. “Effects of Pokémon GO on Physical Activity and Psychological and Social Outcomes: A Systematic Review.” *Journal of Clinical Medicine* 10 (9): 1860. https://doi.org/10.3390/jcm10091860.

Meskell, Lynn. 2015. “Introduction: Globalizing Heritage.” In *Global Heritage*, edited by Lynn Meskell. Wiley Blackwell.

Ohrvik, Ane. 2020. “Den moderne skattejakten: app-teknologi som facilitator for kulturarvliggjøring ved St Olavskilder i Norge.” *Laboratorium för folk och kultur. En kulturvetenskaplig tidskrift* 1: 1–13.

Ohrvik, Ane. 2021. “When Ritual Texts Become Legendary. Practice and Fiction in Nordic Folklore.” In *Fictional Practice: Magic, Narration, and the Power of*

Imagination, edited by Bernd-Christian Otto and Dirk Johannsen. Brill Academic Publishers.
Oring, Elliott. 1986. *Folk Groups and Folklore Genres.* Utah State University Press.
Österlund-Pötzsch, Susanne. 2018. *Gångarter och gångstilar. Rum, rytm och rörelse till fots.* Makadam.
Richardson, Robert C., and William J. Ritchie. 2024. "Geographic Information Systems—A Scavenger Hunt for Accounting and Auditing." *Journal of Emerging Technologies in Accounting* 21 (2): 123–31. https://doi.org/10.2308/JETA-2022-029.
Richter, Felix. 2023. "Charted: There Are More Mobile Phones than People in the World." World Economic Forum, August 11. Accessed December 16, 2024. https://www.weforum.org/agenda/2023/04/charted-there-are-more-phones-than-people-in-the-world/.
Skinner, Heather, David Sarpong, and Gareth R. T. White. 2018. "Meeting the Needs of the Millennials and Generation Z: Gamification in Tourism through Geocaching." *Journal of Tourism Futures* 4 (1): 93–104. https://doi.org/10.1108/JTF-12-2017-0060.
Stolpejakten. Accessed August 26, 2024. https://www.stolpejakten.no/.
Urry, John, and Jonas Larsen. 2011. *The Tourist Gaze 3.0, Theory, Culture & Society* (third edition). Sage Publications Ltd. https://doi.org/10.4135/9781446251904.
Valk, Ülo and Daniel Sävborg. 2018. "Place-Lore, Liminal Storyworld and Ontology of the Supernatural. An introduction." In *Storied and Supernatural Places*, edited by Ülo Valk and Daniel Sävborg. Helsinki: Finnish Literature Society / SKS, pp. 7–24.
Walsham, Alexandra. 2011. *The Reformation of the Landscape: Religion, Identity, and Memory in early modern Britain and Ireland.* Oxford University Press.
Woodward, Matthew. 2024. "Geocache Statistics: Is Geocaching Still Popular in 2024?" *Search Logistics*, June 23. Accessed December 17, 2024. https://www.searchlogistics.com/learn/statistics/geocache-statistics/.

18

More-than-Human Tracks

Walking on the Rendena Cattle Summer Pasture Trails in the Italian Alps

Daniele Valisena

A landscape shaped by traditional cattle farming. This is how Manuel Cosi (2023)—cattle farmer, restaurant owner, and president of ANARE (the National Association of Rendena Cattle Breed Farmers)—described the role of the Rendena cattle breed in Val Rendena, a beautiful valley in the heart of the Trentino Dolomites, in the Italian Alps. Unlike the farming practices typical of the Po Plain, where cattle never leave their stables, in Trentino, some cattle still spend the warm months grazing on the summer pastures in the mountains, retracing centuries-old transhumance trails. In Val Rendena, every spring, a dozen cattle herds are led from their winter stables to the *malghe*—a term that refers to both the pastures and the mountain cottages where the animals and cattle herders live until early September (Raffaetà 2021). This custom attracts thousands of tourists to Pinzolo for the so-called *sfilata*, a week long celebration organized around the return of the cattle from the summer pastures.

Recently, the seasonal movement of livestock has been threatened by economic constraints, overtourism, and the return of large predators such as wolves and bears. Additionally, the push to modernize and scale up the local farming industry raises fundamental questions about development, sustainability, heritage, and biodiversity. Ultimately, what is at stake are the potential futures of the valley and the impact they would have on its socioecological equilibrium.

This chapter aims to investigate the nexus between human and nonhuman mobilities, animal farming, and tourism in Val Rendena. By exploring practices, stories, and mobility patterns, I will examine processes of rewilding, notions of modernization, local and European Union agricultural policies, and the

Figure 18.1 Val Rendena is the north-south valley lying east of Parco Adamello-Brenta and west of Trento. Source: © OpenStreetMap.

various ways in which ideas of sustainability and conservation are understood, implemented, and performed by different actors in the alpine valley. This discussion will build on a series of oral histories and walking ethnographies collected during my two fieldwork visits to Val Rendena between the summer of 2023 and the winter of 2024. In terms of methodology, I use a geohistorical walking method aimed at unfolding the different layers of stories and practices that have shaped the environment over time (Valisena 2022). Drawing on the works of Tim Ingold (2007) on walking and lines, and Baptiste Morizot's concept of tracking (2021), I will reflect on the entanglements between human and nonhuman stories and mobilities in the valley.

Geohistory, Tracking, Lines

According to Antonio Canovi (2011), geohistory is at once a posture, a methodological standpoint, and a theoretical lens. Following Michel de Certeau's (1990) suggestion to adopt the pedestrian view over that of the urban planner, and by focusing on the everyday life practices that constitute lay geographies, Canovi adds a temporal dimension to the urban ethnographer's spatial exploration. Such an approach is not only valid for cities but can be applied to all human geographies. By walking through places, geohistorians immerse themselves in both old and new patterns that have signified those spaces through time, recovering long-forgotten stories and practices that have imbued those sites with different meanings. It is an art of noticing (Tsing 2015), as much as it is an art of moving along and across both visible and forgotten patterns and stories (Van Dooren and Rose 2016).

Building on this epistemological foundation, I have added geological and ecological lenses. Specifically, I seek to uncover invisibilized or forgotten layers of stories, materials, and soils, paying attention to biotas and the ways in which these sedimented strata of the past merge with and traverse present-time stories and ecologies. Walking, in fact, is not merely a process of moving from point A to point B; it is a relational and spatializing bodily act, one that signifies places as a meshwork made of discrete lines that intersect—knots, in Ingold's (2007) term. To navigate this meshwork of trajectories, I find it useful to follow Morizot's (2021) practice of *pistage* (tracking). Tracking is a way to trace and retrace the nonhuman itinerant storying of places and to account for the ways in which non-humans participate in shaping our sense of place and our bodily responses to it.

I conducted this research with my colleague François Thoreau. As a Northern Italian environmental historian from the Po Plain and a Belgian STS researcher—both white, able-bodied males—we were able to access most of the testimonies and sites we visited thanks to the kind and generous intermediation of our colleague Roberta Raffaetà, from Ca' Foscari University in Venice, a resident of the valley. The charm of the Dolomites, combined with the stories Roberta shared, convinced us to embark on this side project to investigate the struggles between rewilding, modernization, sustainability, and traditional ways of life at the frontier between Central and Southern Europe.

Linking Infrastructures, Values, and Movement Heritage

We approached the trail toward Malga Vagliana on a warm early September morning. Leaving the summer house we had rented in Carisolo, we drove to Madonna di Campiglio, intending to explore one of the summer pastures where local cattle breeds from Trentino had spent the last four months. It was just a matter of days before the animals would descend from the mountains back to their stables in the valleys.

Looking at the tourist maps showing the ski lifts and hiking trails radiating from Madonna di Campiglio toward the mountain peaks, we were struck by how these trajectories almost coincided. Hiking trails, ski infrastructure, and transhumance patterns appeared juxtaposed, as though entangled in an unspoken but obvious embrace. Why? Was this the product of chance, or was something else going on—a kind of invisible logic linking these seemingly very different technological, natural, and mobile infrastructures?

In the spring of 2023, the area made the news following the tragic death of Andrea Papi, a twenty-six-year-old local man who died on April 3 after being attacked by the bear Jj4 while running in the woods near the village of Caldes, in Val di Sole. During our first stay in Trentino, the memory of the event—the first of its kind in decades—and the proximity of the provincial elections intensified the debate. Many actors united against the law protecting bears in Italy and the European Union, blaming the Adamello-Brenta Natural Park authorities for reintroducing the animals to the area some thirty years ago (Maestranzi 2023). As it turned out, farmers and cattle herders were also deeply invested in this debate, which touched on themes such as freedom of movement, mountaineers' way of life, human and animal safety, farming, and even transhumance as a practice.

While Madonna di Campiglio was one of the first ski stations to open in the Italian Alps, turning the valley into a popular holiday destination, Val Rendena lacks the agricultural and farming renown of the surrounding valleys. Nonetheless, the efforts of some local entrepreneurs to valorize traditional cheese and cured meat practices have begun to pay off, alongside ski and mountain tourism, which remain the area's primary economic activities. In the wake of this success in tourism, a handful of local cattle farmers have pushed for the revalorization of the Rendena breed. While the breed's origins date back hundreds of years, connecting the valley to Switzerland, zoonosis, and trade routes, the Rendena cattle only gained popularity in the late 1800s, reaching over eighty thousand purebred animals in 1946, farmed across Lombardy, Veneto, and Trentino. However, the state-led modernization of Italian agriculture during the Fascist Era (Valisena 2025) targeted the Rendena breed as a symbol of backwardness. Italian zootechnicians favored Brown Swiss cattle, famous for their milk yield, and considered the Rendena a degeneration of the original Swiss breed. As a result, policies to eradicate the Rendena were enacted in 1931 and again in 1943. Risking arrest by local authorities, farmers resisted by hiding cattle in their stables, ultimately succeeding in preserving the breed. However, agricultural modernization in the 1950s was far more effective in reducing the Rendena population, effectively eradicating most of the breed from the Po Plain and many areas of the Alps (Polla 2014). Today approximately eight thousand Rendena cattle remain, primarily in Veneto and Trentino.

The value of the Rendena cattle breed, inextricably linked to the territory of the valley, is directly connected to the traditional farming practices that have shaped the valley's socioeconomy for centuries. Transhumance and summer pastures are cornerstones of this local ecology. Through our walking exploration, we aimed to retrace and analyze the movement heritage (Svensson et al. 2022) embedded in the transhumance of Rendena cattle, and to examine how it relates to the pressures of tourism and modernization from a wide range of actors. During our time in Trentino, we observed how this debate emerged as a key theme, discussed by stakeholders as varied as unions, dairy cooperatives, zootechnicians, farmers' associations, and politicians. We sought to understand how local farmers reflect on these past, present, and future scenarios, particularly in relation to issues such as climate change, EU and Italian agricultural policy, and new ecological challenges.

Skiscapes, Flight Ways, and Walking Trajectories

The trail we wanted to walk began directly at the meeting point of these various geographies and stories. It was an easy hike, around twelve kilometers there and back, with a difference in altitude of about 580 meters. From below, we could see the ski track cutting an almost straight line through the woods that outlined the slopes of the mountain. Our starting point in Campo Carlo Magno was a crossroads of ski routes, where the Fortini-Pradalago ski lift and the Grostè cable car converge. The grandeur of the ski infrastructure was evident, imposing itself over the landscape like the axes of a giant Cartesian plane, defining the space. In this mid-point stood a hotel with traditional Trentino-Tyrolean wooden architecture, alongside a series of futuristic cottages built in the 1970s and 1980s, all lined up along a big parking lot.

A few cars were parked there, but compared to the winter frenzy and the thousands of skiers that populate the area in that season, it was very calm. Apart from the two of us, there were no other hikers. The contrast in the intensity and volume of tourists across seasons was clear. In the summer, a metamorphosis occurs, turning the slope into a quiet and serene refuge for nature lovers. This observation is significant in relation to the different perceptions of the environment held by human and nonhuman roamers, and the way different actors traverse and inhabit these sites at various times of the year.

Passing the hotel, we went by a golf resort and ventured toward a well-kept prairie sloping up toward the woods. The hike literally followed the ski track, bordering it on one side. While the spatial hegemony exerted by the ski track was uncontested, seasonality also shaped the accessibility—even the visibility—of the trail, which was simply buried under snow in the winter months. As we continued our climb, the trail veered away from the ski track, entering the woods. Spruce and Austrian stone pine trees made up much of the forest. It was quite an ideal Northern European mountain landscape.

The history of the region, which in the past formed the southernmost part of the Austrian-Hungarian Empire, seemed inscribed into the landscape ecology of the area. Deeper into the woods, the path became narrower and more sinuous, following both the ascending inclines and the gaps between the trees. It was evident that the forest had been farmed for several decades at least. It was also easy to spot the areas where the trees had been attacked by the European spruce bark beetle (*Ips typographus*), an insect endemic to European spruce forests that has caused multiple epidemics across the continent (Servizio Foreste e Fauna 2023). Red clusters of dead trees could be seen across the slopes of the valley, reminding visitors and locals that even remote areas like the Brenta Dolomites will not be spared by climate change and global epidemics. The trail was well maintained, as almost everything in Trentino is, and from the tracks on it, we could tell that it was not only frequently walked but also regularly traversed by mountain bikers.

As we ventured deeper into the woods, the distant sound of cars on the provincial road began to fade, replaced by the chirping of birds and the sound of the wind caressing the tree branches. I realized that there were other trajectories intersecting with ours and the ones we were tracing on the ground; the flight ways of local birds also contributed to shaping this ecology, if not the landscape itself. As bipeds walking on the ground, we are accustomed to perceiving and relating to the environments we traverse in a "horizontal way." Our gaze moves from the ground to the sky, from our immediate surroundings to the horizon. We foreshorten the space we inhabit, adapting it to our sensory and motor ways of perception. But what would it mean to account for the ways other living beings move through and make sense of the space they inadvertently share with us humans? How would that shift our sense of place? How might our moving perception attune to those offbeat patterns?

Fence In, Fence Off

After an hour and a half without glimpsing any sign of cattle, we finally came across a yellow nylon cord—a mobile fence. These temporary infrastructures reminded me of the plastic tape used by police to secure a crime scene, though the ones we saw here were much sturdier and more solid in appearance. Stretched between trees, the colored ropes enclosed the area where the cattle could move and graze freely. As I would soon discover, depending on the area and the perceived dangers, these fences could be reinforced with metal wires and electrified. Yet the one we encountered seemed designed mainly to prevent the herd from falling off cliffs or getting lost.

Here was a new layer of mobility—and control over animal movement—beyond those I had envisioned. While transhumance is often understood as a space of freedom for both animals and human herders, the degree of mobility is actually controlled and defined by traditional mobility patterns, as well as by new technological systems of fencing that physically define the perimeter of the animals' freedom. Reducing the grazing space facilitates the herders' control over the animals, but it also impacts the various mobility regimes that intersect with these fences. Younger, more daring cattle can still jump over them—the fences are only about 1 to 1.5 meters tall—while other roaming creatures, including not only human hikers but also deer, wolves, and bears, also encounter these infrastructures of containment.

A couple of days before our hike, we attended a public meeting in Pinzolo. The event was attended by various politicians, farmers' and tourism associations, the director of the Adamello-Brenta Natural Park, and several zootechnicians. Many local farmers were also present. During the public debate, a cattle herder responded heatedly to cautious remarks about the possibility of starting a population management program for large carnivores in Trentino, recounting her personal encounter with a bear. For years, the herder had spent the summer months with cattle herds on summer pastures in the nearby valleys. While occasional predations of young calves and lambs had been rare in previous decades, according to the herder, these incidents had grown exponentially in recent years. The previous summer, she had literally found herself fighting with a bear that had been systematically predating her herd during the night. Under Italian law (Legge 157/1992), "it is forbidden to kill, capture or use violence against protected species," such as brown bears. As a result, the herder had been forced to stay awake day and night to prevent the bear from slaughtering all her animals. No fences, survey cameras, or fire blanks seemed to deter the carnivore, which, over time, had even developed new strategies to bypass the *malga*'s fencing. The herder concluded her story during the public discussion following the conference in Pinzolo (September 7, 2023) by saying, "The experiment to reintroduce the bears was an experiment, right? Experiments are meant to be trials, aren't they? Well, it's clear that this didn't work out well at all, so let's put an end to this."

In his walking ethology of the traces left by wolves and bears, French philosopher Baptiste Morizot has developed the concept of tracking, turning it from a hunter/explorer practice into a transspecies epistemological standpoint. According to Morizot (2021), through this itinerant, embodied practice, attentive trackers can rediscover coevolutionary ways of sensing, living with, and noticing that humans allegedly codeveloped with other species throughout our shared evolutionary history. The French scholar bases his theory on evolutionary biology, arguing that humans have forgotten these arts of living, noticing, and relating due to the decline of such practices and the preeminence granted by Western civilization to sight over all other senses.

Interestingly, while Morizot emphasizes the more-than-human bodily and sensorial skills that humans can rememorate while following the tracks of large carnivores, the account of the herder in Pinzolo suggests the opposite. Animals, too, can learn to read the techno-natural dispositifs put in place by humans. By reading the environment and responding to these dispositifs, animals develop new strategies to breach them, in a way "humanizing" their sense of place and skill set. The capacity of large carnivores to read the environment and overcome human-made impediments emerged repeatedly in our interviews (Polla 2023; Cosi 2024) and informal conversations with farmers in Val Rendena following the meeting in Pinzolo (September 7, 2023). This cognitive ability in animals like wolves and bears is captured in the Italian term *malizia*, which translates as both "malice" and "subtle action," or an "astute and sly behaviour." The subtlety with which large carnivores think, interpret, and move through the environment may represent the greatest threat to the safety infrastructure that has allowed transhumance to exist and to resist the social and economic transformations of the last century.

According to various farmers and other locals, the very freedom to live and move in their territory—a defining feature of mountaineering life—has been compromised by the reappearance of wild animals. The fact that these nonhuman inhabitants of the valley engage in relational and active responses, rather than passive ones, lies at the heart of the problem. These nonhuman inhabitants are living symbols of the fact that domestication is a contingent and historically grounded practice of human control over certain species, and as such, does not govern all ecological relations from the outset (Piazzesi 2024). As a result, many farmers and other locals offer a blunt, peremptory historical reading of the process: wolves and bears were driven to extinction in the Italian Alps for "good reason"—there simply is no space for coexistence.

Walking, Retracing, Entangling

Continuing our climb along the mountain trail, we encountered signs of timbering practices. Historically, timbering was the primary economic activity in the alpine forests. Sawmills, logging, and charcoal production were just some of the activities that occupied locals from fall to spring, sometimes supplementing or even replacing animal farming and transhumance. Pluri-activity and seasonal mobility between the valleys and the plains—or even across the Alps to Germany, Austria, France, or Switzerland—were constituent parts of Trentino's local socioeconomy and mobility ecologies (Grassi 2015). Some of the trails now used by tourists and excursionists are the material remnants of these past timbering patterns, helping to preserve those trajectories and stories through the deep transformations that have shaped the Alps.

Today the economic landscape has changed profoundly. Timbering is no longer the primary activity in the area. Instead, 72 percent of the 390,000 hectares of forest in Trentino (63 percent of the provincial territory) is designated for productive purposes, all managed through a system of sustainable forestry (*silvicultura naturalistica*). Each year, around 6,880 hectares are logged, a practice that allows the woods to regenerate and expand (Servizio Foreste e Fauna del Trentino 2018).

While the total area of wooded land has increased significantly in Trentino over the past century, a number of local actors contest the results of this process. Although indicators in terms of restoring endangered habitats, carbon storage, and the logging industry are all positive, many farmers view the encroachment of forested areas on villages and hamlets in the valleys as a sign of mismanagement. Historically, the lower mountain slopes were cleared of trees and used for agriculture or animal farming—a practice that explains the presence of various *malghe* between the summer and winter pastures on the mountain flanks. Today these structures are often used as summer homes or to accommodate tourists, though some are still used by local farmers as their primary residences and farms. This transformation occurred primarily in the postwar years and before the enactment of the EU-funded LIFE project URSUS, which reintroduced ten bears to the Adamello-Brenta Natural Park between 1999 and 2002.

While the vast majority of actors, who consider the forest environment a cornerstone of their territory and thus of their identity, praise the aesthetic and landscaping value of the forest, the ecological implications of this change in the land are much less appreciated and celebrated. In many mountain communities, rewilding and the increase in woodlands are synonymous with neglect and a sense of loss of earlier agricultural and silvicultural activities that exploited—if not eradicated—parts of the forest. Alessandro Rippa (2024) referred to this as a postworking landscape, underlining the centrality of the now-lost human labor in ordering and shaping liminal alpine environments.

The alpine forest, as an open space of mobility, seems to elude both the biopolitical ordering of space (with humans at the top of the pyramid) and the regime governing the use and accessibility of the territory above the valleys. This is particularly interesting because Trentino represents both a hybrid environment and a meeting point of Germanic and Latin/Roman environmental cultures. Specifically, it is where the traditional Germanic forest cult clashes with the Roman and Italian tendency to view the *selva* with distrust. In Italian tradition, the forest is portrayed as the space of wilderness, in opposition to the realm of civility—the anthropized countryside (Farinelli 2008). Clearing the woods was one of the first things the Romans did whenever they colonized new territories—a key act of renovation and the foundation of a new, civilized space (an ordered environment under the control of *cives*, the citizens of the city). Transforming the environment and its ecology was part of the Roman settlement strategy, which involved constructing roads, building dwellings, and clearing away the "wild" element, whether human or nonhuman.

These cultural tendencies persist in modern Italian environmental culture, though they are now mixed with various other cultural and ecological influences. During the medieval and early modern periods, the clearing and management of forests, as well as their use for communication and work routes connecting the alpine valleys, were constant practices. This daily use made it possible to patrol the woods and selectively control large carnivores, either by reducing their habitats or their numbers. The modernization of the Alps disrupted this bottom-up control mechanism. Specifically, the ways in which people moved through the valleys shifted: from mountain trails to paved roads, from walking to driving, and with a substantial decrease in the rhythms and frequency of forest traversal. A change in the socioecological relationship with the forest, through labor, leisure, and everyday life practices, mirrored a shift in the movement ecology of the Alps.

We could see this change by crossing the tarmac roads now used by locals, forest rangers, lumberjacks, a few tourists heading to their summer homes, and sometimes even cattle farmers moving their animals in what could be defined as mechanized transhumance. Gone are the lumberjack squads camping in the woods for months; the daily trips to collect water, milk, or wood; and the daily movements from home to the fields, or to attend to the various trades that once animated the life of the valleys. New trajectories and itineraries have replaced the old ones, which, in turn, have been taken over by nonhuman inhabitants and roamers.

According to Tim Ingold, all these different trajectories (or threads and lines, as he calls them) are entangled in a meshwork. Drawing upon Henri Lefebvre, Ingold defines these more-than-human meshworks as "the reticular patterns left by animals, both wild and domestic, and by people . . . whose movements weave an environment that is more 'archi-textural' than architectural" (Ingold 2007, 23). While the English anthropologist underlines the value of these itinerant trajectories and practices in weaving together the space, in linking it, what emerged from our walking exploration of this part of the Dolomites was a series of geohistorical aporias—noncorrespondences between previous trajectories and practices of the space and the new socioecological transformations of the environment. The return of former and seemingly forgotten untamed inhabitants of the woods, along with their unruly ways of moving and inhabiting the shared environment of the Dolomite slopes, can be also understood as a form of socioecological forgetting of nonhuman ways of inhabiting the alpine environment.

While my head was spinning with all these thoughts, we arrived at the first stop on our walk: Malga Vaglianella. The building is a long, low, one-story stone structure with a steep gray roof also made of stone. The extensive building was designed to accommodate a large number of cattle—up to fifty or sixty heads—and to prevent heat dispersion. Dominating a small glade where the animals could graze, it was surrounded by a small wooden fence. Though the *malga* looked well maintained and recently renovated, it appeared to be empty. As we found out during our interviews in the valley, tensions had arisen between some of the oldest hut guardians (*malgari*), the municipalities in charge of managing access to the *malghe*, and newcomers who had been renting some of the *malghe* in recent years. These tensions appear to be linked to shortcomings in EU agricultural funding schemes and the customary common law governing access to and use of *malghe* in Trentino.

To summarize, several large farmers from the plains—sometimes using impoverished or complicit local herders as fronts—won the open call to rent out *malghe* as a way to access EU agricultural funding or to justify the excessive dumping of animal waste in the plains (Mencini 2021). These speculative actions are particularly insidious, as they exploit EU liberalization and fair competition policies to infiltrate and profit from centuries-old commons systems regulating the pasturelands in this part of Trentino. In doing so, they threaten the *malga* system, the associated transhumance practices, and even the entire local breed farming economy, which struggles to survive without access to summer grazing lands. Moreover, this issue could force the already dwindling number of farmers in Val Rendena and the surrounding valleys to abandon traditional farming methods and adopt more modern systems, with commercial fodder and breeds. Such a shift would endanger the entire ecological balance that produces and sustains the alpine landscape. It would also impact tourism, as the landscape of places like Val Rendena is a direct result of the socioecological systems that shape it—including, above all, transhumance.

With that in mind, we continued on our way to Malga Vagliana.

The Herd and the *Malga*

Some fifteen to twenty minutes past the first *malga*, we finally encountered some cattle. Or rather, we heard them: the distant sound of bells coming from the trees somewhere behind us. The sound grew clearer, yet we never managed to see the animals or the herders. It was difficult to tell how many were out there, or where they were headed, and on which trail. This was rather disappointing, given our initial goal of following the cattle routes to the summer pastures. However, reflecting on the situation, we realized that a trail as narrow as the one we had been walking on through the woods could hardly accommodate a

Figure 18.2 Malga Vagliana. Source: Photo by Daniele Valisena.

herd of cattle. The animals likely had their own trails—they probably made their own way through the woods, creating their own paths, which they then followed and retraced year after year. This idea was somewhat complicated by the fact that, unlike a few decades ago, not all the animals were raised for dairy or labor anymore. They were also bred for meat now; as a result, for many, this might have been their only season up in the *malghe*. It seemed that the meat and dairy production infrastructure had infiltrated these delicate socioecologies from the plains, dictating new seasonal rhythms and lifespans, thereby disrupting the animals' ability to remember and retrace their own paths to the summer pastures.

After another ten or fifteen minutes, the forest began to open up—first gradually, then abruptly—giving way to a rocky and rolling landscape dotted with sparse young trees. A large glade spread out before us. As the trail continued to climb along the forest's edge, we finally caught sight of a tall wooden cross and three buildings. This was Malga Vagliana, the endpoint of our hike.

Malga Vagliana is a beautiful and sturdy example of traditional summer huts. Its austere appearance reminded me of German huts, with the steep roof almost touching the ground on one side. The upper part of the house was all wood, enveloping the stone body of the house beneath it. Two small windows punctuated the walls of the structure on each side. Gravel and sand delineated the farmyard, where a small outbuilding stood. Behind the main building, an elongated stable—almost identical to the one we had seen at Malga Vaglianella—led to a large concrete watering trough. Four people were sitting in the yard, watching us silently. They were herders—those in charge of tending the animals in the *malghe*. It was a motley crew: a middle-aged man; a young, shirtless blonde man in his early twenties; a boy; and a girl. When they saw us, they began speaking quietly in their dialect, possibly expressing some distrust toward these unexpected visitors—us. After greeting them, we asked if we could fill our water bottles at the fountain. They said yes.

I experienced a certain recalcitrance in questioning them, as I felt deeply foreign to that environment. Furthermore, I was very cautious about addressing the sensitive issue of the pasturelands that had been tearing the herders' community apart in recent years, particularly with regard to EU agricultural funding speculation and semi-illicit grazing practices. Encouraged by my walking companion, I asked the herders a few questions about the animals they looked after. To our dismay, there were fewer than two dozen young bullocks, and none of them were of the Rendena breed. The herders were all from the nearby Val di Sole, and they were looking forward to returning to the valley after four months spent in the summer pastures. That was all they were willing to share with us foreigners. Our presence must have felt like an intrusion on their remote retreat, the umpteenth group of tourists looking for traditional farmers or the closest thing to an "indigeneity experience" that Southern Europe still has to offer. Their whispered exchanges in dialect—a language I can usually grasp a little, due to its similarity to my own dialect, but not in this context—were like a secret code, cutting us out of the life they led, which we were just briefly crossing.

After enjoying a snack and playing a little with a friendly red cat, we decided to head back down. The moment we made the decision to leave, we spotted the herd. It was a small one, with fewer than a dozen animals grazing in the glade that stretched all the way to the large rock that formed the majestic flank of Pietra Grande. We approached them slowly, and for the first time, I realized that free-ranging cows are not necessarily as docile and gentle as my city boy image had led me to believe. When I was about ten meters away from a white and gray cow, she stopped grazing and began looking at me. Her gaze wasn't explicitly hostile, but neither was it a friendly invitation to come and pet her.

She was studying me, while also keeping an eye on the herd and exerting her protective role over the group. Better not to challenge her. I slowly backed away, keeping my gaze fixed on her. She never stopped watching me, while her fellow companions approached her to look at these newcomers dressed in red and black. After a while, when we were about fifty meters away, she returned to her grazing. Walking on the same trails doesn't necessarily mean walking together. Walking together requires a mutual choice, one that both wayfarers agree upon. That didn't seem to be the case here.

While discussing our encounter with my colleague, just a few minutes down the glade, we met the rest of the cattle herd. Our perhaps romantic, or stereotypical, ideas about transhumance had fed us images of shepherds dressed in wool and wearing strange fluffy hats, walking with big sticks. Well, the herders we saw now were the same young guys we had seen before, and they were moving the animals . . . while sitting in a Jeep. About twenty cattle—older and sturdier than the young ones we had encountered earlier—were trotting down the dirty road toward other pastures, maybe the one near Malga Vaglianella.

Figure 18.3 Cattle climbing the summer pasture trail close to Malga Vaglianella. Source: Photo by Daniele Valisena.

These animals were definitely friendlier than the ones we had seen before. They trotted very close to us—though we gave them space—gazing at us. We could tell they were having fun. To our dismay, none of the animals we saw were Rendena cattle. While walking, one might know where to go but can never be certain whom one will meet along the way.

Tourist trails, ski infrastructures, and transhumance patterns all contribute to shaping the socioecologies of the Brenta Dolomites in the Italian region of Trentino. The walk discussed in this chapter was conceived as an itinerant exploration of the various entanglements and frictions that characterize the area. Walking, as a heuristic posture and practice, allows for unexpected encounters and stories to emerge, while adopting a geohistorical perspective helps connect the dots between past, present, and future practices of space—welcoming not only human subjects but also the nonhuman ones traditionally erased from scholarly and anthropocentric accounts.

Spruce plantations and their endemic insect enemies; hikers and skiers; lumberjacks and herders; migratory birds and cattle; wolves and bears; transhumance, skiing, and the geological processes that gave rise to the Dolomites—all contribute to shaping the landscape of this part of the Italian Alps. By crossing these multifaceted and sometimes hidden trajectories, and moving through the stories inscribed within the more-than-human ecology of the Dolomites, it is possible to cultivate "arts of noticing" that could allow for a more nuanced understanding of the ways of moving and inhabiting Trentino and all the Alpine territories. In fact, a geohistorical analysis can allow the attentive wayfarer to see the entanglement of the layered stories that shaped the landscape over time, accounting for human and nonhuman presences in ways that are neither fixed in time nor exclusionary but open to change. In the face of climate change and increasing environmental depletion, learning to navigate ecological transformations in ways that are less anthropocentric and more relational is of crucial importance.

Acknowledgments

The research conducted for this article was funded by the European Research Council (ERC) project "The Body Societal" (grant agreement number 949577), within the Horizon2020 funding scheme. I would like to thank my principal investigator (PI) and fieldwork companion François Thoreau for inviting me to

Trentino and for his enthusiasm. I also wish to express my gratitude to Roberta Raffaetà for her invaluable help in establishing contacts in the valley, for her kind support during our stay in Val Rendena, and for reviewing an earlier version of this chapter. Similarly, I would like to thank Niccolò Martellozzo, Manuel Cosi, Giovan Battista Polla, Valentina Maestranzi, Ivo Povinelli, and Mauro Povinelli for generously sharing their time with us *foresti* and engaging in discussions about the past and the future of their beautiful valley.

References

Canovi, Antonio. 2011. "Pensare nella territorialità, abitare nel paesaggio: introduzione alla Geostoria." Archivio Storico Osvaldo Piacentini. https://archiviopiacentini.it/uncategorized/pensare-nella-territorialita-abitare-nel-paesaggio-introduzione-alla-geostoria/.

de Certeau, Michel de. 1990. *The Practice of Everyday Life*. University of California Press.

Farinelli, Franco. 2008. *L'Invenzione della Terra*. Sellerio.

Grassi, Patrick. 2015. *Sul filo dell'emigrazione*. Centro Studi Judicaria.

Ingold, Tim. 2007. *Lines. A Brief History*. Routledge.

Legge 11 febbraio 1992, no. 157. Norme per la protezione della fauna selvatica omeoterma e per il prelievo venatorio. https://www.normattiva.it/uri-res/N2Ls?urn:nir:stato:legge:1992-02-11;157.

Mencini, Giannandrea. 2021. *Pascoli di Carta. Le Mani sulla Montagna*. Kellerman.

Morizot, Baptiste. 2021. *On the Animal Trail*. Polity Press.

Piazzesi, Benedetta. 2024. *Del Governo degli Animali. Allevamento e Biopolitica*. Quodlibet.

Polla, Giovan Battista. 2014. *La mia vita in . . . e per la Rendena*. Comune di Caderzone Terme.

Raffaetà, Roberta. 2021. "Microbial Antagonism in the Trentino Alps: Negotiating Spacetimes and Ownership through the Production of Raw Milk Cheese in Alpine High Mountain Summer Pastures." *Current Anthropology* 62 (24). https://doi.org/10.1086/714348.

Rippa, Alessandro. 2024. "Where Tourism Fails: The (Un)making of a Post-Working Landscape in the Italian Alps." *Landscape Research* 49 (7): 934–45. https://doi.org/10.1080/01426397.2024.2309310.

Svensson, Daniel, Salzman, Katarina, and Sörlin, Sverker "Movement Heritage and Path Dependence: Layering the Past", in *Pathways: Exploring the route of a Movement Heritage*. edited by Daniel Svensson, Katarina Saltzman, and Sverker Sörlin. White Horse Press.

Tsing, Anna. 2015. *The Mushroom at the End of the World: On the Possibility of Life in Capitalist Ruins*. Princeton University Press.

Valisena, Daniele. 2025. "The Battle of Zootechnics. Incorporating Race, Technology, and Ideology in Cattle Breeding Practices during Fascism in Italy (1922–1945)." *Isis. A Journal of the History of Science Society* 116 (4).

Valisena, Daniele. 2022. "Walking on Terrils. Ruderal Ecologies and Toxic Heritage Wallonia." In *Pathways: Exploring the Routes of a Movement Heritage*, edited by Daniel Svensson, Katarina Saltzman, and Sverker Sörlin. White Horse Press.

Van Dooren, Tom, and Deborah B. Rose. 2016. "Lively Ethnographies: Storying Animist Worlds." *Environmental Humanities* 8 (1): 77–94.

Interviews and Other Sources

Cosi, Manuel. Speech at the conference "La razza Rendena come presidio del territorio montano," September 7, 2023, Palapinzolo, Pinzolo. My translation from the Italian.

Cosi, Manuel. Interview with Daniele Valisena and François Thoreau, Agriturismo La Trisa, Giustino, January 17, 2024.

Maestranzi, Valentina. Interview with Daniele Valisena and François Thoreau, Parco dell'Adamello-Brenta Headquarters, Strembo, September 9, 2023.

Polla, Giovan Battista. Interview with Daniele Valisena and François Thoreau, Maso Pan, Caderzone Terme, September 8, 2023.

Servizio Foreste e Fauna del Trentino. 2018. "Boschi e legno in Trentino." https://forestefauna.provincia.tn.it/Foreste/Foreste-in-Trentino/Le-foreste-in-Trentino.

Servizio Foreste e Fauna del Trentino. 2023 "Piano Bostrico." https://forestefauna.provincia.tn.it/Foreste/Foreste-in-Trentino/Bostrico.

Transcript of the public discussion following the conference "La razza Rendena come presidio del territorio montano," September 7, 2023, Palapinzolo, Pinzolo. My translation from the Italian.

19

Constructing the Path

A Qualitative Exploration on Foot

Hanna Elisabet Åberg

Over the past decade, my parents and their neighbors have expressed concern about a changing landscape—the loss of the landscape as they once knew it. This fear of loss has manifested in, or been shaped by, the establishment of the nature reserve Mullvalds strandskog (Mullvald's Beach Forest) in 2013, nestled in the pine forests of eastern Gotland, Sweden's largest island. Once a community-owned forest along the shoreline, Mullvalds strandskog has been shaped by centuries of pasturing and selective cutting. The distance from communication infrastructures has made it difficult for humans to access. When the reserve was established, the landowners protested, as they no longer had the right to farm the forest, which they claimed to have done for centuries. However, their efforts were without success. Because the forest was close to my family home, I had walked through it from time to time, but to the best of my knowledge, it had never had a name before it became a reserve.

In a time of climate crises, there is perhaps a growing human need to find comfort in nature (Birch et al. 2020), alongside hands-on legislative actions aimed at restoring and conserving natural areas (see, for example, EU 2024, the 2024 nature restoration regulations of the European Union). As a concept, it can be argued that "nature conservation" has developed in parallel with the idea of recreation and nature-based tourism (see, for example, Spirn 1996; Mels 2002). Drawing on examples from northern Sweden, human geographer Tom Mels (2002) contends that nature-based tourism can serve as a tool for promoting a territory, further emphasizing the importance and uniqueness of the place. As Gotland has seen a significant increase in tourism over the last decade, its unique nature has become a trademark of the island, with Mullvalds strandskog nature reserve now part of a network of natural areas transformed into destinations. The status of "destination" adds importance and value to what is found within the borders of such sites.

Just like the borders that define Mullvalds strandskog nature reserve, the idea that a line on a map determines what is valuable and what is not has been scrutinized by numerous academics over the past decades (see, for example, Corner 1999). Critical heritage scholar Christina Fredengren (2015) claims that the protection of natural areas and the sustainable integration of cultural heritage practices have long been subjects of academic discussion. However, scholars within heritage studies have been relatively slow to embrace the nonhuman turn in thinking, according to both Fredengren (2015) and heritage scholar Colin Sterling (2020). The wedge between nature and culture has long been a feature of Swedish landscape conservation ideology, particularly emphasized in the 1960s and 1970s following a reorganization of the Swedish environmental administration, and it remains evident at the national level today (Germundsson 2005). Fredengren (2015) argues that heritage policies continue to place people at the center. In environmental policy, nature has been treated in a similar way to environmental protection (Bennett et al. 2016). This human-centered thinking in policymaking and planning continues to create a divide between nature and culture, challenging the "sustainable" aspect of sustainable landscape planning when establishing reserves like Mullvalds strandskog.

Let us adjust and redefine the meaning of "landscape" in sustainable landscape planning. The concept of landscape has been broken down and scrutinized by various disciplines to the point that it is, by definition, ambiguous (Burlingame 2024). It has served, and perhaps still serves, as a mediator in the nature-culture discourse, feeding into the posthuman rhetoric of today. Policies and conventions (for example, the Cultural Landscape Convention and the European Landscape Convention [ELC]), are attempts at natureculture integration, although they still privilege human perception (Olwig and Lowenthal 2006). The very definition of landscape, as stated in the ELC—"Landscape means an area, as perceived by people, whose character is the result of the action and interaction of natural and/or human factors"—positions it within an intangible heritage discourse, according to Fredengren (2015). However, this ambiguity in the concept of landscape could encourage more direct interventions in informing policy (see Bartolini and DeSilvey 2021). Ensuring biodiversity protection in sustainable landscape planning should not overlook the human importance of a sense of place (Devine-Wright 2009), as collective identity is often anchored in physical spaces (Rosenberg 2012).

This chapter aims to explore contemporary nature conservation and sustainability in landscape planning from a critical heritage perspective through walking. It engages with the work of Christina Fredengren (2015), who calls for a heritage approach to sustainable policymaking that transcends natureculture, and architect Elissa Rosenberg's (2012) reflections on walking in relation to identity. I conducted two ethnographic walks in the spring of 2018 and 2024. In 2018, I concluded two informal talks with former landowners (in May) and reviewed official documents related to Mullvalds strandskog (Länsstyrelsen 2013; 2016; 2018). By cross-referencing the results of these different methods, this chapter presents a historical account of the landscape of Mullvalds strandskog that is empirically grounded and open to interpretation in a broader social context, particularly in relation to the national Swedish planning discourse. The aim is to encompass different conservation perspectives and record material change rather than to criticize the process of reserve creation. In this chapter, I will take the reader on two walks to Mullvalds strandskog, reflecting on identity, sustainable landscape planning, and, of course, on the inevitable act of walking itself.

Walking as a Method

In a recent essay, geographer Matthew Gandy (2024) calls for attentive observation through walking fieldwork to encounter multisensory and multispecies textures. Indeed, we perceive our immediate surroundings with all our senses (Rantala et al. 2020). Coming from a scholarly practice in architecture, urban planning, and heritage, I have been taught to ease walking, to explore by walking, and to design walking. In urban planning, walking methods are used to (re)connect with place. It is believed that the stories of a place unfold as the individual moves, feeding into a cohesive understanding of that space (Cantillon 2019), raising awareness of almost forgotten stories, encouraging a meaningful sense of place (Markwell 2007), and assembling ecological knowledge and identity (Anderson 2004).

Architect Elissa Rosenberg (2012) defines different practices of walking from an architectural perspective: walking as an act that makes us see the world differently, walking as an everyday practice, and, finally, walking as a journey in itself. In Rosenburg's example, walking as a journey is exemplified by a memorial that uses movement to create the space of commemoration. While walking, layers of time and connectedness to the surroundings seem to emerge. Yet, Rosenburg argues, place is constructed by the walker; it is neither stable nor reliable, and narratives are not fixed to be read but open to interpretation.

The first walk described in this chapter was never intended to be rewalked. Reflecting on it now, this walk was conceived as a journey in itself (as Rosenberg describes it), where the movement aimed to unpack the layers I had experienced through reading conversation and management plans, as well as informally speaking with former landowners. These layers were stories of use, material changes, and long lists of valuable species. In line with de Certeau's weariness suggestion (1988), I tried then (and again now) to employ the pedestrian perspective over my own, the urban planner's, while exploring the landscape of Mullvalds strandskog. Coming back years later, the walk itself was the journey once again, but with an emphasis on attentive observation, reenacting the embodied experience of having walked this path before—this time with the intent to record landscape change.

Reading the Landscape of Mullvalds Strandskog

Mullvalds strandskog is located about an hour from Visby, the only town on the island, in what most people would consider a remote area. Situated two kilometers from the road, the beach forest consists of sixty-four hectares of old mixed coniferous forest, sand pine forest, and various types of sand dunes. It grows mainly in drifting sand and is one of the most extensive and well-preserved dune sand heaths on Gotland, according to the official management plan. The natural values include the open dunes and the old tree layer of spruce and pine, which provide an important habitat for a variety of unusual insects, soil fungi, and plants. Embedded in the forest is a relatively large formation of rauks, unusually far from the shore. The hard reef limestone from which the rauks are formed tells of a time when Gotland was a marine environment at the equator, about 440 million years ago. This particular rauk, known as the "King's Hall," gets its name from its round formation, which makes it look like a fortress.

In 2012, the Swedish government decided to protect 150,000 hectares of forest in Sweden. By 2020, 4,300 hectares of forest had been protected on Gotland, which is 1.4 percent of the island's total landmass. As a result, Mullvalds strandskog was declared a nature reserve in 2013. Becoming a reserve meant that landowners lost ownership but received economic compensation based on the value of the forest. The landowners protested, claiming that they and their ancestors were responsible for the area's high natural values (Sveriges radio 2013). From their perspective, access to the forest would be limited, and they openly disagreed with how local authorities managed their forests (Sveriges radio 2013). A protection plan was established at the end of 2016, stating that the ecological values were a result of careful human management. In addition to these natural values, the management and establishment plan emphasized the high social values of the area, especially in terms of recreation and tourism. As a result, the management plan (Länsstyrelsen i Gotlands län 2016) proposed building a road and a car park. In 2020, several local development companies initiated and established a longer hiking path across the area (Sveriges radio 2020). That same year, new recreational infrastructure was put in place, such as benches and barbecue facilities. Since 2022, a yearly trail run has been organized along the route.

In the following sections, I take two walks through Mullvalds strandskog, six years apart.

Wandering Mullvalds Strandskog in 2018

When I visited in 2018, there was no official road leading up to the reserve. The two kilometers from where the tarmac road ended seemed endless—could it be that there were more potholes than solid ground? There was no parking, so I left the car by the side of the road. From there, I began my walk.

I was not sure where the reserve started. What were the boundaries? Geographer Kenneth R Olwig (2008) writes, "The power of the map and the landscape scene is greatly enhanced when the national boundary is not just a line on paper, but also marked by a clear, 'natural' division of the elements, especially that between land and water" (79). As an island, Gotland's outer borders are indeed natural. Water automatically legitimizes a nation or a region as "natural." As an island, Gotland has a strong regional identity shaped by a common language, history, values, and geography. Especially from an external perspective, the Gotlandic identity could be interpreted as it is expressed through the regionalization of nature, as seen in advertisements or travel guides (see, for example, Länsstyrelsen i Gotlands Län 2018). The Gotlandic landscape is often associated with the sea, the coasts, and the barren land (called alvar). However, the more recent focus on high-value forests suggests a shift in how the island's landscape is viewed.

As I walk from the car through Mullvalds strandskog nature reserve, I think about Olwig's words. The reserve is not a large area, but it has no natural boundaries, as the forest extends without clear limits. The woods seem to go on forever. With my limited ecological knowledge, I can't tell the difference between what is valuable—the area where I stand in the reserve or what I see a few hundred meters away. The trails are narrow and few, and there is no map to guide me. I have the feeling that I have lost my way.

I wander around for quite a while, nearly losing faith in finding the "Hall of the Realm" or "King's Hall" (Figure 19.1) (personally, I prefer the latter name, as that is how it is known locally). A guide to the area explains that the lush spruce forest and the inland rauks create an enchanted environment (Länsstyrelsen i Gotlands Län 2018). When I finally find the right path, a narrow one, I am reminded of Rosenberg's description of walking as a journey: "[A] path is designed as a journey or passage to a meditative space of commemoration, invoking a sense of the sacred or sublime" (2012, 135). I have the sense that nature has designed

Figure 19.1 "The Hall of the Realm" or "King's Hall." Source: Photo by Hanna Elisabet Åberg, 2018.

a grand entry to a sacred abode. The tall, dense trees shade the ground, and the sun's rays—it is spring, after all—dramatically hit the strange stone formations.

In 2018, Gotland County Council published a travel guide to Gotland's nature, describing Mullvalds strandskog as one of the island's most valuable forests that has developed freely over a long period of time (Länsstyrelsen i Gotlands Län 2018). In fact, however, most of the forest has been maintained, used, and managed by local farmers for centuries. There is a tradition of selecting trees to be felled, and the area has also been used for grazing throughout history. In the village, a long-established rumor is often repeated, not unlike other local tales: one farmer allegedly felled part of the forest on the edge of another reserve out of sheer fear of also losing that land. He believed that once the forest was cleared, it would lose its biodiversity value and thus become unattractive from an environmental perspective. This story perhaps reflects a deeper concern among landowners—not a fear of losing income but of seeing their decades of work, management, knowledge, and expertise of this particular place go unrecognized. Instead of the landowners continuing to manage the land, an external team would now implement a management plan—one that concludes that trees will likely have to be removed to preserve the forest "as it is."

As I leave the forest behind me and wander down the sand dunes, the surroundings seem to perfectly reflect commercial images of Gotland: an empty beach and an endless blue sea. The management plan suggests removing the wild roses (*Rosa rugosa*) from the beach, as they are considered invasive and a threat to other species (Länsstyrelsen i Gotlands Län 2018). It is too early in the year for the beauty of their flowers, but their thorny presence, in the form of shrubs, is apparent.

I walk back to the car, not having met a single human soul during my walk. I think about the characterization of Mullvalds strandskog as an enchanted environment. I felt it too, according to my notes. Also wanting it to remain as it was, I silently find myself agreeing with the management plan.

Walking the Path of Mullvalds Strandskog in 2024

About six years later, I visited the reserve again. On the way, my parents' neighbor, a former landowner, tells me that the dirt road is too bad to use. Since my last visit, a gravel road was built, but, according to the neighbor, the pressure of car use earlier that summer has made it unusable. I'm not sure if this is true or if it's the result of years of conflict between the former landowners and the regional authorities. Although curious to see the road, I decide to take the neighbor's advice and drive down a nearby road, from where I have to walk a few kilometers before entering the reserve.

In recent years, a trail has been built along the coast that follows the coastline for several kilometers. As I walk along this path into the reserve, I pause to consider the best way to cross a river. It's filled with spring water now, and it's hard to find a place to cross. Stopping here makes me think of de Certeau (1988) and his idea of walking as a strategic discourse, not merely an activity with a goal. He argues that while walking, people make calculated decisions—whether to pass by, cross, create shortcuts and detours, or forbid others from taking a path. As I contemplate the water, three mountain bikers pass me, riding fast and fearlessly across the river. I'm taken aback by their boldness and regret not saying hello to them afterward. After a beautifully orchestrated jump—a strategic decision—across the river, I soon find my way into the reserve.

Over the years, I have heard former landowners talk about the degradation of the forest. Trees that have fallen of their own accord are left untouched. The trunks are good for certain species. Is this planning for nonhumans, I wonder? How can humans carry the burden for all? My thoughts wander as I enter the reserve. I think again about Olwig's (2008) concept of "natural" boundaries, as they seem to have emerged as I approach the reserve. The forest is much less dense than the one behind me, with fallen trunks scattered here and there. An official sign directs me to the "King's Hall (Figure 19.2)." I didn't need the sign, I realize, as the forest was so open that I could see the formation straight from the path. Although I recognize the "King's Hall" (Figure 19.2) unmistakably, that wordless sensation—a sense of magic or some other unworldly emotion—has changed. Maybe it's the influence of the former landowners, maybe it's the gray weather, maybe it's a material change, or maybe it's a personal result of aging and having gathered more experiences and visual impressions. However, the feeling of being surrounded by greenery, with tall trees and cracks of light—how I had previously described and remembered it—is no longer there.

There are three barbecue areas within the reserve, along with four information points and two car parks. There is also a real road, though it's rather worn (Figure 19.3). Following the road past the largest car park to the south, several modern houses have popped up just outside the reserve's borders. I assume they are second homes. With the new infrastructure and the houses looming through the trees, that feeling of perhaps losing my way is no longer there.

Figure 19.2 “The Hall of the Realm” or “King’s Hall." Source: Photo by Hanna Elisabet Åberg, 2024.

Figure 19.3 A barbecue area on the beach. Source: Photo by Hanna Elisabet Åberg, 2024.

There are a few people along the path, and I have a short chat with some of them about the weather, the rauks, and a fallen tree trunk—material aspects of nature can be discussed from endless viewpoints (see Midholm and Saltzman 2014). Wandering off the road toward the shore, I even bump into an old friend. We've never been here together before. She happens to be a spatial planner just like me. Also like me, she is here with a special intention—though totally unrelated to our common profession: she is training for a marathon. She parked her car a few kilometers east and ran toward the reserve on the map. This makes me think of the idea of escapism—entering nature to escape your everyday life. And yet here is a part of my everyday life: talking about marathons.

Conclusion

Coming back here, years later and with new knowledge, I reflect on the understanding of knowledge. The fear of loss and landscape change that my parents and their neighbors express is tied to their anchored identity. Their memory is an active, participatory practice that shapes their present (see Rosenberg 2012), grounded in their knowledge of place. There is a common understanding that expertise is crucial when it comes to solving environmental problems. While experts are often understood as academic professionals or (in this case) authorities, other types of knowledge holders are also critical for local resilience, according to environmental sociologists Lidskog and Berg (2022). However, experience-based expertise is difficult to evaluate and demonstrate and can be challenging for outsiders to recognize and understand (Lidskog and Berg 2022). Environmental issues are complex, and an important part of the regulatory process is to consider not only environmental aspects but also what people value and how they are likely to respond to decisions.

In accordance with Rosenberg's notions (2012), the outcomes of these visits show that places are not stable entities but important anchors for identity. While these ideas are not new, they are worth emphasizing: a better understanding of place-based identity, both at the individual and societal levels (Devine-Wright 2009), may be essential when planning and establishing a new reserve. Theoretically, a new reserve like Mullvalds strandskog would have ecocentric aspirations. The natural infrastructure that enables certain biotopes may not be visible to my untrained eye. However, picnic benches, new paths, and general wear and tear are. I can't help but wonder how ecocentric aspirations can result in such anthropocentric outcomes.

Through my two walks in Mullvalds strandskog and my conversations with former landowners, I learned that the loss of land remains an open wound. This isn't because of the economic value of the land but due to the lack of trust in their knowledge and good intentions. Olwig and Lowenthal (2006) have shown how natural heritage can be socially and culturally constructed, often serving national agendas. The establishment of one relatively small nature reserve contributes to these larger national and international goals, though it is a mere drop in the bucket of what is needed by international standards (EU 2024). May (2019) points out that heritage is often understood as elements of a shared human inheritance preserved for the benefit of future generations. However, it is often unclear how professionals can make decisions in the present that are truly in the best interests of all. If we are to truly aspire to protecting a shared human heritage, it is better to involve others in the process. In a posthumanist era that transcends naturecultures, protection efforts must embrace new ways of thinking, adapting, and promoting biodiversity, while also aligning with societal values and ethics (Svenning et al. 2024).

What about walking as a method? What emerged from my two walks was not a comparative study of material change but rather an exploration of sense of place. As Rosenburg puts it, the layers of time and connection to the environment became visible through an embodied memory evoked by physically retracing my own footsteps and later rereading my notes. The place—then and now—is my construction as a walker. Recovering my footsteps revived memories of the senses, embodied ones, stored within me in shades of sharp spring light and a collage of green, beige wet sand and the soft tones of bark. As I walked, I remembered without thinking.

The establishment of Mullvalds strandskog as a nature reserve has created a local social wound and a mistrust of authority among some residents. The developed planning and infrastructure in Mullvalds strandskog, visible to the human eye, seem designed for (human) visitors. It is planning for future generations of visitors rather than for those who are rooted here. Learning from this, integrating local knowledge, and gaining a better understanding of identity are valuable in all conservation efforts and key to discussing sustainable landscape planning.

References

Anderson, Jon. 2004. "Talking Whilst Walking: A Geographical Archaeology of Knowledge." *Area* 36 (3): 254–61. https://doi.org/10.1111/j.0004-0894.2004.00222.x.

Bartolini, Nadia, and Caitlin DeSilvey. 2021. "Landscape Futures: Decision-Making in Uncertain Times, a Literature Review." *Landscape Research* 46 (1): 8–24. https://doi.org/10.1080/01426397.2020.1861228.

Bennett, Nathan J., Robin Roth, Sarah C. Klain, et al. 2016. "Conservation Social Science: Understanding and Integrating Human Dimensions to Improve Conservation." *Biological Conservation* 205: 93–108. https://doi.org/10.1016/j.biocon.2016.10.006.

Birch, Jo, Clare Rishbeth, and Sarah R. Payne. 2020. "Nature Doesn't Judge You—How Urban Nature Supports Young People's Mental Health and Wellbeing in a Diverse UK City." *Health & Place* 62 (2): 102296. https://doi.org/10.1016/j.healthplace.2020.102296.

Burlingame, Katherine. 2024. "Landscape Geographies: Interdisciplinary Landscape Research and a New Framework to Apply Landscape as Method." *Landscape Research*: 1–12. https://doi.org/10.1080/01426397.2024.2369682.

Cantillon, Zelmarie. 2019. "Urban Heritage Walks in a Rapidly Changing City: Tensions between Preservation and Development on the Gold Coast, Australia." *Journal of Heritage Tourism* 15(2): 149–63. https://doi.org/10.1080/1743873X.2019.1618315.

Corner, James. 1999. "The Agency of Mapping: Speculation, Critique and Invention." In *Mappings*, edited by Denis Cosgrove. Reaktion Books.

de Certeau, Michel. 1988. *The Practice of Everyday Life*. Translated by Steven Rendall. University of California Press.

Devine-Wright, Patrick. 2009. "Rethinking NIMBYism: The Role of Place Attachment and Place Identity in Explaining Place-Protective Action." *Journal of Community & Applied Social Psychology* 19 (6): 426–41.

European Union, Regulation (EU) 2024/1991 of the European Parliament and of the Council of 24 June 2024 on nature restoration and amending Regulation (EU) 2022/869, (2024), ISSN 1977-0677.

Fredengren, Christina. 2015. "Nature: Cultures—Heritage, Sustainability and Feminist Posthumanism." *Current Swedish Archaeology* 23 (1): 109–30.

Gandy, Matthew. 2024. "Attentive Observation: Walking, Listening, Staying Put." *Annals of the American Association of Geographers* 114 (7): 1386–404. https://doi.org/10.1080/24694452.2024.2353841.

Germundsson, Tomas. 2005. "Regional Cultural Heritage Versus National Heritage in Scania's Disputed National Landscape." *International Journal of Heritage Studies* 11 (1): 21–37. https://doi.org/10.1080/13527250500036791.

Länsstyrelsen i Gotlands Län. 2013. *Skötselplan för naturreservatet Mullvalds strandskog.* Visby: Länsstyrelsen i Gotlands Län. https://www.lansstyrelsen.se/download/18.299c1cd2167c07c647510e2/1545211223628/Mullvalds%20strandskog_skötselplan.pdf.

Länsstyrelsen i Gotlands Län. 2016. *Bevarandeplan för Mullvalds strandskog.* https://www.lansstyrelsen.se/download/18.2e0f9f621636c84402717de8/1527682589149/Mullvalds%20strandskog%20SE0340133.pdf.

Länsstyrelsen i Gotlands Län. 2018. *Utflyktsguide: 45 besöksmål i Gotlands natur.* https://catalog.lansstyrelsen.se/store/30/resource/81.

Lidskog, Rolf, and Monika Berg. 2022. "Expertise, Lay/Local Knowledge and the Environment." In *Handbook of Critical Environmental Politics*, edited by Luigi Pellizzoni, Emanuele Leonardi, and Viviana Asara. Edward Elgar Publishing.

Markwell, Kevin, Deborah Stevenson, and David Rowe. 2007. "Footsteps and Memories: Interpreting an Australian Urban Landscape through Thematic Walking Tours." *International Journal of Heritage Studies* 10(5): 457–473. https://doi.org/10.1080/1352725042000299063.

May, Sarah. 2019. "Heritage, Endangerment and Participation: Alternative Futures in the Lake District." *International Journal of Heritage Studies* 26 (1): 71–86. https://doi.org/10.1080/13527258.2019.1620827.

Mels T. 2002. "Nature, Home, and Scenery: the Official Spatialities of Swedish National Parks. *Environment and Planning D: Society and Space* 20(2): 135–154.

Midholm, Lina, and Katarina Saltzman. 2014. *Naturen för Mig: Nutida Röster och Kulturella Perspektiv.* Institutet för språk och folkminnen.

Olwig, Kenneth R. 2008. "'Natural' Landscapes in the Representation of National Identity." In *The Ashgate Research Companion to Heritage and Identity*, edited by Brian Graham and Peter Howard. Routledge.

Olwig, Kenneth R., and David Lowenthal. 2006. *The Nature of Cultural Heritage and the Culture of Natural Heritage: Northern Perspectives on a Contested Patrimony.* Routledge.

Rantala, O., Salmela, T., Valtonen, A., & Höckert, E. 2020. "Envisioning Tourism and Proximity after the Anthropocene". *Sustainability*, 12(10), 3948. https://doi.org/10.3390/su12103948

Rosenberg, Elissa. 2012. "Walking in the City: Memory and Place." *The Journal of Architecture* 17 (1): 131–49. https://doi.org/10.1080/13602365.2012.659914.

Spirn, Anne Whiston. 1996. "Constructing Nature: The Legacy of Frederick Law Olmsted." In *Uncommon Ground: Toward Reinventing Nature*, edited by Willian Cronon. W. W. Norton & Company.

Sterling, Colin. 2020. "Critical Heritage and the Posthumanities: Problems and Prospects." *International Journal of Heritage Studies* 26 (11): 1029–46. https://doi.org/10.1080/13527258.2020.1715464.

Svenning, Jens-Christian, Melodie A. McGeoch, Signe Normand, Alejandro Ordonez, and Felix Riede. 2024. "Navigating Ecological Novelty Towards Planetary Stewardship: Challenges and Opportunities in Biodiversity Dynamics in a Transforming Biosphere." *Philosophical Transactions of the Royal Society B* 379 (1902). https://doi.org/10.1098/rstb.2023.0008.

Sveriges radio. 12 September 2013. https://sverigesradio.se/artikel/5642998.

Sveriges radio. 24 June 2020. https://sverigesradio.se/artikel/7501934.

20

Powered by Nature's Call

Fecal Footprints and Biopolitics in a Nordic Tourist Landscape

Rose Keller and Roger Norum

Nature-based tourism (NBT) is a dynamic and swiftly growing industry, deeply intertwined with the climate crisis. The debates it raises can be a powerful force for both environmental and social change. NBT refers to tourism based on the observation and appreciation of natural areas for leisure. In NBT, the individual desire for experience is coproduced from deep-seated human needs (such as connection, stress reduction, and positive emotions) and is facilitated by strategic marketing and narrative generation in which imagery is often "inspired by"—or appropriated from—nature. Places like national parks can be key drivers of NBT due to their economic value, serving as a "marketplace" for nature experiences. However, this has begun to wear on the conservation values of intact ecosystems (Keller et al. 2025).

This chapter traces a story of conservation that unfolds through a novel applied research project aimed at addressing one of the increasingly unavoidable byproducts of NBT's expansion into remote nature: human fecal accumulation. Led by the first author, "Traceless Tourism" engaged visitors in adopting a new hiking practice—packing out their own feces using a waste-alleviation and gelling (WAG) bag—in the highly visited Lofotodden National Park in Norway. However, the walking of purposeful transects to engage with fecal traces left behind revealed ingrained notions of human-nature relations, as well as beliefs surrounding control, markets, and authority.

In Norway, the rise of NBT as a viable economic activity in a country historically dependent on fossil fuel revenues has led to a rampant increase in tourism commercialization and "arctification"—a process that reifies (and packages) northern nature as a space where the imaginaries of a remote and

Figure 20.1 Kvalvika Beach, study area and midnight sun viewing tourist hotspot. Source: Photo by Rose Keller.

pristine landscape can be experienced up close and in person (Rantala et al. 2019). The consequences of this growth have been noted by scholars for several decades (see, for example, Kaltenborn et al. 2019). However, in recent years, tensions have emerged due to the increasing number of tourists and the expansion of NBT activities, which have outpaced the capacities of park managers and local communities (Fossgard and Fredman 2021; Haukeland et al. 2023). National parks in Norway are typically established to protect the natural qualities of a place and its various aspects (for example, certain species). But as Norway becomes more reliant on tourism income, these parks are also increasingly compelled to attract tourists with outstanding recreation opportunities. Given that establishing infrastructure to support tourism in remote areas (and under different models of environmental governance) can be challenging, park managers often prefer solutions that shift or regulate visitor practices to encourage the adoption of new "desired" behaviors that promote leaving a more conscientious tread and reduce the detrimental impacts of high use.

One particular leave-behind of the tourism experience is currently terraforming the land, reconstituting human-environment relations and place-

Figure 20.2 Biffybag (human waste portable bag), open and ready for use. Source: Photo by Rose Keller.

based dynamics: feces. As NBT expands, the problem of open human defecation in protected and natural recreation areas is growing globally (Marion et al. 2016). In many nature-based destinations, an increasing number of visitors leave not just footprints but also fecal remains. This chapter focuses on how tourists leave such traces and how they relate to them: a biopolitics of tourism's excrement. We ask what the implications are for NBT when "natural" practices, unalienable from the human body and its multiple mobilities and mobile acts, are considered undesirable or inappropriate for a given place. How do actors (that is, authorities, tourists, entrepreneurs) conceive of and address the less sanitary aspects of being human when faced with an increase in outdoor recreation opportunities? And what effect does this growth have on natural landscapes, such as national parks? We are particularly interested in what happens when individuals produce and encounter waste (their own and that of others) in spaces where natural areas become crowded, trammeled, and sullied. The imposition of asking tourists to think about—and sometimes handle—their own waste can conflict with how resources are earmarked for visitor management, creating concrete ramifications for the approaches and practices of NBT.

In the following pages, we outline the project's problems, tactics, and results before discussing the paradoxes of conserving pristine nature by unpacking individual subjectivities and internalized control through ethnographic reflections with the project's participants, using Michel Foucault's biopolitics framework (1990; 2008).

Waste Is a Terrible Thing to Mind: Defecation and Mitigation via Intervention in Lofotodden National Park

The problem of toilets and human waste in tourism landscapes has been taken up in recent years by a number of scholars in tourism studies and related fields (see, for example, Nevelyn and Aruan 2020; Mensah et al. 2023). It is an interesting issue to reflect upon, perhaps less due to experiential concerns (Will tourists come to a place if they have to worry about their bodily waste?) or planning challenges (How should a tourism destination plan sanitation if visitor numbers increase?) but rather for its existential implications (How are bodies and nature governed, for what purpose, and by whom?).

Indeed, the problem of open defecation in protected and natural recreation areas is not merely a challenge of mitigating awkward encounters in nature. It is also a biophysical and health issue, one that is increasingly important as so-called last chance tourism (Lemelin and Whipp 2019) to remote areas rises globally. Outdoor recreationists of varying expertise travel to experience pristine nature and to "escape reminders of modern society" (Kaltenborn et al.

2019), such as basic sanitation infrastructure. This contributes to the growing challenge of surficial human waste, pollution of local water bodies and soil, and the potential spread of disease. This issue is most readily observed when visitors deposit their feces, toilet paper, menstrual products, and other waste on or just below the surface of the land (or snow cover), where it can be dug out by animals or eventually seep into and contaminate the landscape and waters (Apollo 2017) through nutrient loading and the introduction of new microorganisms (Scott et al. 2022). When surface water is subsequently used for recreation, drinking, or food preparation, it can contribute to the spread of bacteria such as *E. coli*, viruses, and protozoa (Cilimburg). Other animals' lives can be impacted in different ways by human byproducts, as they consume microplastics, hormones, antibiotics, or even become addicted to psychoactive drugs like caffeine (Baron et al. 2023).

This problem of human waste and trash is not limited to visible lowland areas but is also a growing health concern in highland and mountain regions such as the climbing routes of Mount Everest in Nepal (Neupane 2023) and Denali National Park in Alaska (Ewert 1991), where mountaineers frequently experience intestinal infections (Bishop 2015) and gastroenteritis (Apollo 2017). Authorities in Lofotodden have also recorded visitors reporting gastroenteritis after ingesting water from open streams in popular camping areas nearby (Keller and Engen 2022; Keller et al. 2023). In what follows, we outline the team's project intervention and discuss its primary aims.

Lofotodden National Park is located on the southwest side of the Lofoten archipelago off the central Norwegian coast, often referred to as *yttersida* ("the outskirts") by locals. Lofoten is one of Norway's most popular destinations, and its national park attracts many visitors. The park's most visited area is Kvalvika beach, a 0.5-kilometer-long camping area accessible via a hike from two parking spots—one in Torsfjorden and another in Innersanden—where "vanlifers" and other independent or guided groups park their vehicles. In order to reach the beach, hikers follow a marked trail for two kilometers. About a quarter of the daily visitors set up tents along the line of vegetation that stretches from west to east along the beach, often to observe the midnight sun. From May to October 2023, the beach saw approximately fifty-two thousand visitors, according to automatic trail counters, a figure that includes guided groups of up to thirty people, based on observations by the "Traceless Tourism" team. A baseline visitor survey conducted in 2019 found that 81 percent of visitors to Kvalvika were first-time visitors (Lindøe 2022). In subsequent years (2021 and 2022), this proportion either remained steady or increased, with a high percentage (87 percent) of nondomestic visitors (Keller et al. 2023). According to the survey, 34 percent of respondents stayed overnight. As a result, Kvalvika beach transforms each night into a sort of "small town" formed by these makeshift campsites. However, this "town" lacks a sewage system; the nearby streams become places not only for relieving one's thirst but also for relieving oneself. These water bodies offer some of the few depressed areas along the beach, making them attractive "hidey-holes" for doing one's business. Waste accumulation is a significant management concern, not only due to high visitor density but also because of the prevalence of coastal tundra and dune vegetation—Lofotodden's shallow, sandy, anaerobic soils have low recovery and decomposition rates for fecal material and toilet paper.

Through "Traceless Tourism," we sought to address the issue of waste buildup in three specific phases. First, we mapped, on foot, the extent and rate of feces accumulation at Kvalvika. We recorded the physical movements and flows of visitors to identify the areas where they were most likely camping and accessing the beach. We also established a regular water quality testing regime in the beach waters, with a control site in a stream away from visitors. Second, we collected and analyzed the attitudes, perceptions, and practices of visitors regarding outdoor defecation, using on-site surveys and short interviews conducted over two years (2022–2023). Last, we introduced and tested a new waste practice for visitors: packing out their own feces using a portable toilet bag. WAG bags are a common individual-level strategy imposed on visitors by managers of parks and climbing routes in remote settings where toilets cannot be maintained. These bags typically consist of an inner and outer bag: the individual excretes into the inner bag, which is then sealed within the outer one. The bags contain chemicals to initiate the decomposition process, dry out the substance, and render it "stink free." Visitors are expected to carry their used bags until they reach an appropriate disposal container.

The goal of the intervention was to influence tourists' attitudes and practices by reinforcing environmental consciousness, drawing attention to both the unique and vulnerable coastal tundra landscape and the bacterial risks (for example, *E. coli*) posed by the unnatural accumulation of feces on the beach. We tested two delivery methods: placing an informational sign prominently at the start of the trail (a passive, minimal-engagement intervention) and speaking personally with each incoming visitor (an active intervention). Outdoor recreation literature generally agrees on the effectiveness of personal communication over passive methods (Selvaag et al. 2023), and by testing both approaches, we could evaluate any variation in effectiveness. We did this based on the existing governance model in Norway, anticipating that parks would be limited to establishing signs, given that personal contact requires more staffing and resources.

To measure accumulation and ensure no spillover effects from the different intervention phases, we cleaned the beach and allowed one full day with no treatment between the two communication phases. Each toilet waste deposit was counted at the end of each control and communication intervention and compared with overnight visitor numbers to generate an outcome variable of average waste per capita. Additionally, we provided clearly marked WAG disposal containers in two convenient locations.

The results of our communication interventions clearly demonstrate effectiveness in reducing the accumulation of human feces and improving water quality at Kvalvika. General visitor characteristics, such as age, gender, group size, and residence, did not differ significantly between the control and communication intervention periods, meaning any change in feces quantity can be attributed to the type of communication (or lack thereof) visitors received. Statistical analysis showed that both passive and personal communication interventions significantly reduced waste accumulation compared to the control periods. The largest and most significant improvement was observed when comparing the control (pre- and postbaseline conditions) to the personal communication condition, which proved to be the most effective, culminating in a 34 percent reduction in waste per capita (Keller et al. 2023). We also measured a significant drop in *E. coli* levels.

Wasted Reflections: Walking and the Biopolitics of Managing the Tourist Body

Emerging from Foucault's examinations of the neoliberal state and forms of power, a biopolitical lens highlights the political production of life, driven by a desire to manage populations through values, language, institutions, and the very conceptualization of individuals' life and death (Lapointe 2024). The control of individual mobility (Iaquinto 2024) or bodily functions is connected to discipline (Minca 2009), the contradictions between freedom and appropriate ways of being, the subjectification of individuals to market or social norms, and the place-making processes of appropriation (Ek and Hultman 2008). In this analysis, we examine several core themes that emerged from interviews with participants (N=76). Using a biopolitical framework, we bring these together by identifying three main themes: control, contradictions in "leave no trace," and the subjectification of the tourist role and body.

Control

Walking with tourists and interviewing them after handing them a bag to poop into was an interesting experience to explore notions of self-control versus control of others. In this context, "control" refers to instances of internalized norms—those things that individuals feel they should do on their own. In some situations, conversation among interviewees served to mitigate responses, such as during a follow-up intervention (see Norum and Keller, this volume), where walkers quietly discussed with a neighbor what they observed as they both scanned the ground. Control appeared when visitors spoke about the need for others to be more educated, better behaved, or restricted in some way. When infrastructure, like toilet facilities, was mentioned, it was often associated with a sense of control: controlling others' behavior through an ordered system that both curbs others' "bad" behavior and dictates their visual experience of the landscape. However, other visitors argued that adding more facilities might attract even more people, which could be detrimental, voicing concerns that this infrastructure could erode the sense of freedom. One individual emphasized the importance of open access without payment for facilities: "[W]e are not used to this concept of paying for any kind of facilities in parks, and I guess that is sometimes good, but many non-Norwegians wouldn't think that's true, because it's supposed to be this concept of nature for everyone. Which I think is really nice about Norway. It's so open" (Interview 6). Finally, some viewed the suggestion of using bags as "overreach" (control by the government) and bemoaned their tainted nature experience: "Making people use these things . . . it's like . . . being told you're a kid and back in diapers. We're hiking, not supposed to be thinking like this" (Interview 31).

Contradictions in "Leave No Trace"

Participants highlighted social norms closely connected to the idea that "modern humans do not belong or behave in nature." Many participants expressed a desire to shield nature from human interference through restrictions on individual mobility. Somewhat counterintuitively, many stated that as long as human infrastructure or impacts remained invisible, nature appeared to remain untouched—making human presence feel acceptable. Some Nordic participants expressed a sense of possessing superior knowledge about Nordic nature yet paradoxically practiced personal waste habits that contradicted the principles of "leave no trace," which they believed were inherent to the Nordic way of being outdoors. This paradox suggests a sense of exclusion, where individuals advocate for the preservation of nature from human intervention but simultaneously feel entitled to access and use it.

Perhaps unsurprisingly, many nature tourists expressed a desire for exclusive ways of "being in nature." Protecting nature for future generations (of humans) was a clear part of this narrative, and using the bags was seen by some as a way to "offset" their present impact: "The bags are a bit like . . . helping me feel less guilty about traveling here" (Interview 44). Some participants strongly believed in excluding those who did not align with their expressed environmental attitudes, which contradicts the narrative of allowing others to experience nature: "[B]ut if you're going to behave badly—and there are so many dumb people out there who just don't care—you have to keep those ones out!" (Interview 39). There is also something deeply contradictory about the "leave no trace" principle itself: it promotes seeking to "touch" nature without actually "touching" it. One participant explained: "A close feeling of nature is what I want, and I don't want to see that others leave things in nature. It reminds me that it's not untouched. Those are not common anymore—untouched experiences" (Interview 51).

Quantity, density, and frequency of visitation all contribute to the impacts on a place, and "leave no trace" represents a set of norms meant to shield nature from human use. While participants agreed that everyone should take out what they bring in (the "leave no trace" principle), their eco-ethos generally stopped there. The theme of contradictions in "leave no trace" is closely related to how individuals perceive their roles in nature: for some, nature should be visited but not imprinted, while for others, it should be an open arena for all forms of acting in nature.

Subjectification

This term locates a subject's practices of boundary management and perceptions of the different roles they perform in different contexts. "Leave no trace" is a set of ordered practices regarding one's role in nature, but it often treats humans as being distinct from natural beings. This distinction breaks down when we feel the need to defecate outside. Tourists tended to perceive themselves as natural beings—part of nature—when defecating in nature: "Humans are not part of nature . . . but then again, pooping is natural, part of nature" (Interview 43). Visitors sought to "acquire the authentic experience of nature" (Interview 67) while expressing concern about the diminishing quality of pristine nature as tourism grows. They subjectified their own bodies as problems yet did not hold the broader tourism industry responsible for increasing demand on resources.

Another element of subjectification that arose in our interviews was the idea of performing the undesirable. Participants expressed the notion that feces should be expelled, continuing the management of the boundary between self and other, or self and outside (nature). As Susan Morrison (2015) observes, this is analogous to equating oneself with refuse. The processes and practices of defecation are thus carefully governed by ordered relations to avoid associating excretion with the self: excreting in nature disrupts this order, forcing people to defecate outside of socially ordered behavior. The awkwardness surrounding fecal encounters in tourism experiences highlights the intersection of subjectivity and the control and discipline of bodies. While the latter may seem somewhat obvious, as society has long established culturally appropriate rules (that is, norms) for the right way to relieve oneself, the subjectivity of this process is less clear. Though fecal matter and related topics are universally understood in terms of disgust, shame, and dirtiness (Lévi-Strauss 1964), the act of defecation is in fact stigmatized to varying degrees based on culture and gender. Martin Weinberg and Colin Williams (2005), in their study of navigating bodily functions within trans communities, draw on stigma theory to show how men are more socially permitted to be open about their bodily functions, whereas women performing feminized roles tend to experience greater constriction. As one female participant put it: "To shit out in the open, without cover or privacy—it feels really uncomfortable. So I hold it" (Interview 52). This discomfort around mundane bodily functions in a context of a sublime nature experience led some female participants to refrain from overnight trips due to the prospect of having to defecate outside of socially ordered bodily norms: "I never would use a bag, but it is because I never stay overnight so I can do *it* [poo] properly [using a toilet]" (Interview 13); "We never do anything else than day trips, and plan accordingly, so we never have to *go* [poo] outside" (Interview 38). In this sense, the tourist body is subject to its own and others' excrement. Contending with this could even be stigmatizing, shifting the interview conversation away from the uncomfortable issue of redrawing body boundaries and back to the ordered relations of gender and cultural norms.

The preceding themes, which more or less cover the arc of the interviews, naturally overlap, particularly when it comes to the overarching issue of "ownership." Specifically, who owns human waste, who owns order in nature, and who should be responsible for both? Ownership is closely tied to biopolitical governance, as it defines the boundaries of responsibility and control. While walking with participants, it became clear that the practice of walking prompted individual reflections on their own personal impact on Kvalvika. However, it also revealed a tendency to skirt the notion that their own walking and waste production were part of a larger collective issue. In other research from this project, visitors were categorized into specific "compliance types," based on their willingness and reasons for adopting the new environmental practice of using a WAG bag. Upon reflection, compliance type seems closely related to the actual type of walking we used to engage intervention participants. Some walks were closer in pace to a run, making it difficult to discuss the more complex issues of waste management and human-environment entanglements. Other walks were more leisurely and saw participants proudly displaying their (newly collected or used) WAG bags—some even with a bit of a swagger. For some, walking with their own waste and discussing it openly was a new experience, and on several occasions, even an enlightening one.

In sublime nature, as opposed to the mundane urban daily grind, NBT can be seen as a form of liberation, evoking assumptions of achieving higher emotional states, such as feeling exceptional (Jaffari 1987) or privileged to experience pristine nature that few others have access to (Salazar and Graburn 2014). However, the compulsion to find a place in nature to relieve oneself can shatter this sought-after, elevated experience: "I came here to seek nature, now all I can think of are these damn shit bags" (Interview 70). While some participants were able to preserve their higher-plane experience of nature by framing the "shit bag" experience within the themes of "leave no trace" or control (that is, appealing to logics of care and altruism), for others, it reminded them of their drab, quotidian lives and needs. However, survey results from this project revealed that seeing toilets was less desirable and more disruptive to the nature experience than seeing a bin with WAG bags.

Participants we interviewed largely expressed remorse about their (necessary) actions as defecating tourists, reflecting an internalized disconnect between what nature "should be" and what it becomes when encountered in real life. An important—and, for participants, markedly disturbing—part of the study was the "Traceless Tourism" team's responsibility for cleaning up human waste between various communications phases to map differences in waste accumulation. For the first cleanup, we recruited volunteers from the local council, the park board, and Visit Lofoten, the region's destination management organization (DMO). The volunteers' interactions began somewhat jokingly, which we understood as an attempt to alleviate the discomfort of facing the task at hand: cleaning up feces in nature that their organizations advertised as pristine and untouched.

By the end of the long day of wrinkled noses, scraping shovels, and heaving sighs, the group looked somewhat forlorn. Against the backdrop of the fjord, the lead author asked for reflections as everyone walked toward the freedom from carrying waste promised by the car park ahead. As one DMO employee put it, "We didn't know this was such a problem, but with luck, your work will convince people to change their ways a little when visiting." While not entirely unexpected, this response was a bit disappointing, as it revealed industry representatives' belief that they do not share the same responsibility as individual tourists. Contradictions can often emerge in ascribing responsibility: participants reported that the main responsibility for dealing with waste lay with (other) tourists, while simultaneously acknowledging that the high number of visitors was largely due to the effective marketing and word-of-mouth accessibility of Nordic nature. The national parks in Norway are required to work with tourism industry representatives to market parks as destinations, which effectively shifts the authority for determining the balance between conservation, visitation, and traditional use away from public representatives. This shift in responsibility from managing supply (marketing) to responding to demand (tourists paying for experiences) reflects a strong internalized "rule" and exemplifies the biopower of the industry-state: over 70 percent of visitors in the "Traceless Tourism" survey (N=556) reported that the main responsibility for fecal and other consequences of visitation lies with tourists themselves.

The long-term effects are likely to differ from the initial impact of this project. In February 2024, the lead author convened a workshop with local stakeholders to collectively discuss the costs and benefits of implementing a bag delivery system in Lofotodden and other areas of Lofoten's highly prized and beautiful nature. Ultimately, the group we worked with decided to invest in an additional two thousand WAG bags, which were distributed for visitors' use in summer 2024. The national park provided free bags at the same locations as in 2023 but did not offer a "park greeter" as we had done during our intervention, due to limited resources and the fact that only one person manages the park. Some local guiding businesses charged 75 NOK per bag to the few clients who chose to purchase them. According to the park manager, the amount of waste reduction was much lower in 2024 compared to 2023, a result complicated by a 13 percent increase in visitors: "Visitors were very willing to take these bags, but without someone there, it is impossible to supply enough or get them to the people who need them most" (Kvalshaug, personal communication 2024). When free, the bags are readily taken, and some visitors use them, often motivated by an informational sign or other visitors. However, long-term solutions must involve the collaboration of both locals and the industry to fund and maintain active interventions that establish and root new environmental practices, tailored to the local context.

Wasting Wilderness: Concluding Thoughts Toward an Affirmative Biopolitics

In this final section, we summarize our findings by drawing connections between the concept of biopolitics and the lessons learned from various components of the Lofotodden intervention. At the individual level, biopolitics can be understood as an affirmative means of transforming society, where tourists are seen not as victims but as empowered subjects capable of building or recovering self-awareness. As Michael Hardt and Antonio Negri (2009) stress, the formation of subjectivity in biopolitics lies in the resistance to and negotiation with biopower. This is where possibilities emerge. In our intervention, participants reflected on the affective states of walking a transect, which ranged from disgust and sadness to encouragement, as passersby engaged with and expressed interest in the activity, results, and consequences of human (in)action.

Another consideration that emerged is the notion of protected areas becoming enclosures that restrict large number of visitors. Some national parks have been curated as "elite" destinations, a strategy designed to limit numbers and impact (Gurung et al. 2022). However, this practice risks turning nature into a space of exception based on systems of exclusion and territorial recoding (Lapointe 2024). Moreover, aggregate impact often remains the same, as luxury tourism generally demands more services and access (Koot et al. 2024). We argue, instead, that conservation policy should focus on mobilizing concepts like vitality and conviviality, by establishing restrictions on the mass use of areas, such as through a lottery system (see, for example, Fix et al. 2012) if necessary. More importantly, it requires furnishing public authorities with the support structures needed to provide either concrete individual-level tools or larger infrastructures to cope with the constant pressures placed on conservation values.

However, this must be approached with a critical eye. Developing the "vitality" of rural societies—in regional development and maintaining stable populations—as a core value delivered though tourism is something that DMOs and interest groups often promise, and municipal planners tend to repeat (Fossgard and Fredman 2021). In order words, this approach risks protecting nature through consumption (in the form of tourism). Instead vitality and conviviality can be recast in local conservation efforts by focusing on relationship building—providing support for public waste discussions and guides to help visitors. Rosi Braidotti (2019) sees vitalist potential in civil society, and in this sense, NBT represents a form of biopolitical governance, or what Erik Swyngedouw (2023) describes as the restructuring of identity and desire, where tourists' autonomy and responsible character and the management of the self are emphasized. Empowering voices in civil society is essential for transforming deterministic biopolitics into an affirming biopolitics that focuses on the life-giving, vital entanglements between humans and nature—distinct from consumption and transactional relationships. Indeed, "leave no trace" comprises a set of practices that align with the individual-based, experience-driven economy of NBT, as it tends to frame environmental action around individual mitigations rather than encouraging individuals to consider the density and extent of use. In the status quo, WAG bags could be seen as an intervention that prolongs the transactional relationship between tourists and tourism, recoding and terraforming places. But under an affirmative biopolitical frame, WAG bags could be understood as opportunities to invest in local volunteers, to reframe parks as spaces of entanglement rather than transaction (that is, the delivery of a particular promised nature-experience), and to (re)build individual nature connections—a metabolic cycle coming full circle.

As we have described in this chapter, walking with waste may take different paths, but the link from a footprint to waste remains the same. What drives increased foot traffic and unfortunate consequences in some Norwegian national parks is tourism, coupled with the decentralized and underfunded public governance of nature conservation. The intervention in Lofotodden exemplified the consequences of regions turning almost entirely to the international tourism industry—one in which the free natural beauty of a region is appropriated to increase tourism, based on the quality of its conservation values, while simultaneously degrading the images and narrative of pristine nature and waters. Such interventions can help highlight the consequences of high use before they are made manifest in the landscape or in people's connection to nature. Walking participants through a given scenario—real or imagined—can reveal how quickly or extensively real and dangerous contaminants and contagions can occur. For participants, this walk brought to the fore notions of wilderness, what it means to be or feel free, and the constant demands placed on communities to facilitate the appropriation of nature, as well as on public authorities to assume appropriate stewardship of it.

Thinking critically about the experience of excrement—whether on an individual level or in terms of present or future growth—can illuminate for visitors, residents, and policymakers the extent to which tourism can penetrate the capillaries of the living world, including our own intimate natural processes. Have we come so far that we now demand that both individuals and nature cope with their own natural waste rather than having the industry pay for the externalized costs associated with selling "free" nature experiences?

The processing of human waste is typically taboo. But if the burden is shared, and human waste can be destigmatized by removing the invisibility of its accumulation and decomposition—and if we bring the topic into scientific inquiry and thought—it is also possible to process it more conscientiously and effectively. If concrete environmental aims are indeed desirable, then a deliberate and unabashed look into our own feces—and the often shitty business of undermining public participation in the conservation of nature—will likely be necessary.

References

Apollo, Michal. 2017. "The Good, the Bad and the Ugly—Three Approaches to Management of Human Waste in a High-Mountain Environment." *International Journal of Environmental Studies* 74 (1): 129–58.

Baron, Jill, Tim Weinmann, Varun Acharya, Caitlin Charlton, Koren Nydick, and Scott Esser. 2023. "Marmots Do Not Drink Coffee: Human Urine Contributions to the Nitrogen Budget of a Popular National Park Destination." *Ecosphere* 14 (4). 10.1002/ecs2.4504.

Bishop, Brent. 2015. "Peak Poop: The Feces Problem on Everest Needs a Solution." *Outside*, April 7. https://www.outsideonline.com/outdoor-adventure/climbing/peak-poop-feces-problem-everest-needs-solution/.

Braidotti, Rosi. 2019. "A Theoretical Framework for the Critical Posthumanities." *Theory, Culture & Society* 36 (6): 31–61.
Cilimburg, Amy, Christopher Monz, and Sharon Kehoe. 2000. "Wildland Recreation and Human Waste: A Review of Problems, Practices, and Concerns." *Environmental Management* 25: 587–98.
Ek, Richard, and Johan Hultman. 2008. "Sticky Landscapes and Smooth Experiences: The Biopower of Tourism Mobilities in the Öresund Region." *Mobilities* 3 (2): 223–42.
Ewert, Alan. 1991. "Managing a Scarce Natural Resource: The High Altitude Mountaineering Setting." Paper presented at the International Association for Experiential Education, Lake Junaluska, North Carolina, October 24–27.
Fix, Peter J., Andrew Ackerman, and Ginny Fay. 2012. "Estimating Visits to Denali National Park and Preserve: Spring/Summer 2011." Natural Resource Technical Report NPS/AKR/NRTR—2012/641. National Park Service, Fort Collins, Colorado.
Fossgard, Knut, and Peter Fredman. 2021. "The Nature-Based Tourism Product." In *Nordic Perspectives on Nature-Based Tourism*, edited by Peter Fredman and Jan Vidar. Edward Elgar Publishing.
Foucault, Michel. 1990. *The History of Sexuality: An Introduction. Volume I.* Translated by Robert Hurley. Vintage.
Foucault, Michel. 2008. *The Birth of Biopolitics: Lectures at the Collège De France, 1978–1979.* Translated by Graham Burchell. Picador Palgrave Macmillan.
Gurung, Deep Jyoti, Paridhi Brahma, and Chandan Goswami. 2022. "Sustainable Luxury Tourism: Promises and Perils." In *The Emerald Handbook of Luxury Management for Hospitality and Tourism*, edited by Anupama S. Kotur and Saurabh Kumar Dixit. Emerald Publishing Limited.
Hardt, Michael, and Antonio Negri. 2009. *Commonwealth*. Harvard University Press.
Haukeland, Jan Vidar, Peter Fredman, Liisa Tyrväinen, Dominik Siegrist, and Kreg Lindberg. 2023. "Prospects for Nature-Based Tourism: Identifying Trends with Commercial Potential." *Journal of Ecotourism* 12 (4): 1–18.
Hawkins, Gay. 2006. *The Ethics of Waste: How We Relate to Rubbish*. Rowman & Littlefield.
Iaquinto, Ben. 2024. "Backpacker Tourism: Definitions, Methods, Debates." *Tourism Geographies*: 1–9.
Jafari, Jafar. 1987. "Tourism Models: The Sociocultural Aspects." *Tourism Management* 8 (2): 151–59.
Kaltenborn, Bjørn Petter, and John D. Linnell. 2019. "Who Are Legitimate Stakeholders? National and Local Perceptions of Environmental Change in the Lofoten Islands, Norway." *Polar Geography* 42 (4): 236–52.
Keller, Rose, and Sigrid Engen. 2022. "Perceptions of Human Waste Mitigation in Kvalvika, Lofotodden National Park: A Mixed Method Study of Visitor Impacts,

Behaviors, and Waste Mitigation Preferences." NINA Report 2374. Norwegian Institute for Nature Research.
Keller, Rose, and Sigrid Engen. 2022–2023. Traceless Tourism Interview Database. Norwegian Institute for Nature Research.
Keller, Rose, Sigrid Engen, and Sofie Selvaag. 2023. "Testing Visitor Management Strategies to Reduce Human Waste in a Highly Visited National Park in Norway." NINA Report 2385. Norwegian Institute for Nature Research.
Keller, Rose, Sofie Selvaag, Sigrid Engen, and Ole-Jakob Kvalshaug. 2025. "Testing Communication Strategies for Reducing Toilet Waste in a Norwegian Protected Area." *Journal of Interpretation Research* 30(1): 119–148. 10925872251325551.
Koot, Stasja, Bram Büscher, and Lerato Thakholi. 2024. "The New Green Apartheid? Race, Capital and Logics of Enclosure in South Africa's Wildlife Economy." *Environment and Planning E: Nature and Space* 7 (1): 123–40.
Lapointe, Dominic. 2024. "The Tourism Periphery: From Structural Hierarchies of Place to Relational Ontology." *Tourism Geographies* 26 (6): 1–10.
Lemelin, Harvey, and Paul Whipp. 2019. "Last Chance Tourism: A Decade in Review." In *Handbook of Globalisation and Tourism*, edited by Dallen J. Timothy. Edward Elgar.
Lévi-Strauss, Claude. 1964. *The Raw and the Cooked: Introduction to a Science of Mythology, Vol. 1*. Translated by John and Doreen Weightman. Harper Colophon.
Lindøe H. E. 2022. *Exploring challenges of visitor-generated waste in Lofotodden National Park* [Master's thesis, Norwegian University of Life Sciences].
Marion, Jeffrey L., Yu-Fai Leung, Holly Eagleston, and Kaitlin Burroughs. 2016. "A Review and Synthesis of Recreation Ecology Research Findings on Visitor Impacts to Wilderness and Protected Natural Areas." *Journal of Forestry* 114 (3): 352–62.
Mensah, Justice, Benjamin Yaw Tachie, and Harriet M. D. Potakey. 2023. "Open Defecation Near a World Heritage Site: Causes and Implication for Sustainable Tourism and Heritage Management." *Journal of Cultural Heritage Management and Sustainable Development* 13 (1): 167–84.
Minca, Claudio. 2009. "The Island: Work, Tourism and the Biopolitical." *Tourist Studies* 9 (2): 88–108.
Morrison, Susan Signe. 2015. *The Literature of Waste: Material Ecopoetics and Ethical Matter*. Palgrave Macmillan.
Neupane, Sangeeta. 2023. "Mountain Tourism and Sustainability: The Dark Story of Everest Trash in Nepal." Master's thesis, University of the Algarve.
Nevelyn, Josephine, and Daniel T. H. Aruan. 2020. "'Toilet Matters': Qualitative Study About the Importance of Toilet in Decision Making in the Context of Nature-Based Tourism." In *Proceedings of the International Conference on Business and Management Research (ICBMR 2020)*. Atlantis Press.
Rantala, Outi, Suzanne de la Barre, Brynhild Granås, et al. 2019. *Arctic Tourism in Times of Change: Seasonality*. Nordic Council of Ministers.

Salazar, Noel B., and Nelson H. H. Graburn. 2014. *Tourism Imaginaries: Anthropological Approaches*. Berghahn.

Scott, Laura C., Scott M. Esser, Alexandra Aubee, Nicholas Lee, Matthew T. Flood, and Tiong Gim Aw. 2022. "Use of Microbial Fecal Indicator Monitoring and Fecal Source Tracking as a Park Management Tool in Rocky Mountain National Park." *Environmental Challenges* 8: 1–9.

Selvaag, Sofie K., Rose Keller, Øystein Aas, Vegard Gundersen, and Frode T. Singsaas. 2023. "On-Site Communication Measures as a Tool in Outdoor Recreation Management: A Systematic Map." *Environmental Evidence* 12 (1): 14.

Swyngedouw, Erik. 2023. "Capital's Natures: A Critique of (Urban) Political Ecology." In *Turning Up the Heat*, edited by Maria Kaika, Roger Keil, Tait Mandler, and Yannis Tzaninis. Manchester University Press.

Weinberg, Martin S., and Colin J. Williams. 2005. "Fecal Matters: Habitus, Embodiments, and Deviance." *Social Problems* 52 (3): 315–36.

21

Esoteric Limping

A Speculation on Supernatural Walking

Ami Skånberg

Figure 21.1 Please scan the QR code to listen to the instructions.

Please scan the QR code to listen to instructions. You can also read them here:

> *Please stand up if possible.*
> *Feel your feet grounded in the floor.*
> *Slowly shift your weight to the right.*
> *Look at your environment while leaning toward the right.*
> *Slowly shift your weight to the left.*
> *Look at your environment while leaning toward the left.*
> *Please repeat this until you find a comfortable rhythm.*
> *Imagine that you are influencing the stars as you sway.*
> *Please continue reading this text while swaying or have a seat.*
>
> (Skånberg 2024a)

In my research, I explore walking as an artistic and choreographic movement practice, examining both its functional and meditational qualities as well as its connections to storytelling and mythologies. This text, situated within the field of artistic research and walking as artistic practice, invites you to engage with swaying movements while prompting questions about their origins and their effects. How did these movements come to be? How do they affect us here and now? What histories are embedded within them? Tending to society and history through choreography and movement supports nonverbal research and the embodiment of historiography.

Since 2000, I have practiced traditional Japanese dance and theater in Japan. My PhD research focused on the walking step *suriashi*, essential in Japanese performing arts, Sumo wrestling, tea ceremonies, and martial arts (Dahlstedt 2022). I also presented a workshop on *suriashi* for the One by Walking research group in Vålådalen (Skånberg 2022b). Historically, there is a connection between *suriashi* and the swaying step I discuss and describe in this text. In my search for explanations of swaying movements in spiritual dances, I discovered a connection between the ceremonies of Japanese Kagura (神楽) (entertainment for the gods) and a Daoist walking ritual (Lancashire 2001). In the Daoist ritual, "the effect of certain stars, considered potentially harmful, was neutralized through walking and foot stamping movements" (Lancashire 2001, 36). The Daoist walking ritual is, in turn, linked to Emperor Yu, which I explore further shortly.

Reflecting on the fact that these choreographed walking styles once served specific societal functions—such as purifying spaces, neutralizing harmful cosmic forces, and bringing peace to the planet—raises an important question: How can we assign similar meanings to the arts in contemporary society, especially in these unruly times? What if, instead of security guards, artists were tasked with safeguarding spaces—keeping them open and playful rather than closed and fearful?

Emperor Yu's Walking Step

The focal point of this text is Yubu (禹步), Emperor Yu's walking step. Yubu is a swaying walk from right to left, through which the walker seeks a connection to the stars. In the legends, Emperor Yu is presented as the founder of the early Xia dynasty (c. 2070–1600 BC). His mythological status invites speculation about the story itself. The myth of Emperor Yu and his continuous walking practices is especially intriguing for research on embodiment, choreography, and choreographed walking. Revisiting this mythology reminds us to value practices that may seem too illogical, unnecessary, or ghostly within our contemporary context.

Moreover, swaying movements exist in multiple cultural contexts, each carrying its own meaning. How might these movements be connected across different traditions? I have gathered various narratives surrounding swaying and aim to share historical accounts of this movement logic, alongside peaceful choreographic acts designed to create awareness and care within society. Experiencing and practicing the swaying step—moving from right to left—brought up three points of interest. First, the narratives surrounding the swaying step itself, and how it came to be, provide an interesting exploration of how new movement strategies emerge and how people might begin to repeat them for ritualistic or artistic purposes. Second, the societal rules—as we know them—about what constitutes "correct" forms of walking, and how new movement strategies challenge and reshape our perceptions of what constitutes "correct walking" also merit examination. Third, in considering the aesthetics of beauty and the adjustments made to transform something perceived as unworthy into something of real value, the question arises: What kind of social change is necessary to bring about this transformation?

I propose "esoteric limping" as a contemporary interpretation of Yu's swaying. The term was coined for my presentation at One by Walking's seminar on March 4, 2021, held on Zoom. In my research, I explore the story behind what drove Emperor Yu to develop his swaying walking style.

Walk with me to learn more.

The Swaying Body and Ritual

The swaying movements and our experiences of them have a potential that we cannot fully grasp, however swaying and rocking together are common movements in many rituals. These movements are practiced in a number of social dances, such as the Japanese Bon festival, disco, jazz, hip-hop, and others. The Bon festival is a celebration for the spirits of the dead, where people dance with their ancestors. It is "a time of pleasure and of quiet recollection of dear ones gone, and also a beautiful event symbolizing continuity with the infinite world beyond" (Ashikaga 1950, 228). When dancing at the Bon festival with many others on a warm Tokyo summer night in July 2024, I sensed how dancing, swaying, and walking together created a strong sense of community that transcended age and language. Older Bon dances blended with newer ones, and participants helped each other remember the movements. Thus, dancing with our ancestors at the Bon festival was not a hallucination; it was a very real practice performed by everyone in contemporary Japanese society. While moving, we created a space where humans, nonhumans, and ghosts could coexist. The essence of this logic also exists in Japanese performing arts, including Butoh (舞踏)—an art form that emerged in 1959 through the collaborations of its two key founders, Tatsumi Hijikata and Kazuo Ohno, and is known to for its resistance to fixity.

There are other examples where rocking plays an essential role, such as among the women in Zár cultures of Egypt, Ethiopia, and northern Sudan. Here rocking represents a response to chanting spirits (Boddy 1989, 161), allowing for interaction with the spirit world, primarily for healing. The body functions as both a vessel and a medium for receiving spiritual messages, with the rocking movement acting as a bridge between the material and spiritual worlds. Zár dancing is practiced by a diverse group of people—Muslims, Christians, and Jews around the Red Sea, the Arabian Sea, and the Arabian Gulf (El Guindy Howaida and Claire Schmais 1994; El Hadidi 2016). This highlights the cross-cultural and interreligious nature of this practice, showing how similar movements and rituals can be shared across different faiths. It suggests that, while the purpose of the ritual may be healing, the act of swaying or rocking transcends specific religious identities. It becomes a universal human experience, in which body movement plays a significant role in spiritual well-being. Another example is the rocking to and fro in the men-only Zikr ritual of the Islamic Sufi tradition, which is seen as a way to gain spiritual enlightenment and achieve union or annihilation in God (Boddy 1989, 129).

In both Zár and Zikr, the rocking and swaying play a central role in rituals where spirits are called upon, helping to bring about a transcendent experience of Oneness. These rituals remind us of the interdependency between the physical body, the spirit, and the community. They show how movement can serve as a medium for processing the complexities of the human experience within which spaces, humans, and nonhumans are cared for and transformed.

In addition, the swaying, playful movements in the Chinese martial art Tai Ji (太極) are practiced and recognized globally. I, myself, studied with Shao Lin monk Zhong Bang Chen in Gothenburg, Sweden. In Tai Ji, the careful meditational weight shifting from right to left is known for its therapeutic effects, but it originally stems from self-defense practices. Tai Ji embodies the notion that all things in nature evolve through dynamic, interactive dualities—yin and yang (Li 2014, 9). Rooted in nature's innate capacity for balance, its swaying movements become a dance of equilibrium, lingering delicately between opposing forces. I contend that this gentle swaying opens a space for ambiguity, challenging the need for rigid, straight lines and embracing fluidity instead.

Sway with me to Beijing.

Emperor Yu, the Swaying Step Yubu 禹步, and Mythology

While attending the World Philosophy Congress in Beijing in 2018, I had the opportunity to interview Wang Xue (王雪) (interview with author, August 19, 2018), a postgraduate student and staff member at the Beijing Dance Academy. She shared with me the story behind the swaying walk in China that is said to have originated with Emperor Yu. Professional dancers, walkers, and shamans continue to perform this step, known as known as Yubu (禹步), to calm the universe. Xue explained that a record of Emperor Yu's step can be found as early as the fourth century in the esoteric book *Baopuzi* by the scholar Ge Hong (283–343 AD) (Hong, Ge. Ca. fourth century). She confirmed that the Yubu is incorporated into Tai Ji (太極). Xue also showed me several pictures of the Big Dipper (北斗七星) and a notation for the walk, performed to calm the seven stars that form the Big Dipper (Ningning 2009, translated by Wang Xue). Finally, she read from *The History of Ancient Chinese Music and Dance*, explaining how Emperor Yu's Walk came to be: "It has been recorded that Emperor Yu had a leg disease from when he attempted to govern a flood, so that he can no longer walk like common people. Individuals call his step 'Yubu,' and it has spread, especially through supernatural dances" (Ningning 2009, translated by Wang Xue).

The story of the Great Yu is deeply embedded in Chinese creation mythology and continues to be passed down through generations. Art historian Heping Lui notes that the new ruling elite of Confucian scholar-officials often cited Yu, famed for his upright morality, to justify arguments and strategies for flood control (Liu, Heping 2012, 92). Yu's flood control efforts became a metaphor for governing the empire, with engineering and innovation celebrated as tools for maintaining order. Working tirelessly, Great Yu built dikes and dams along the riverbanks, creating a system of irrigation canals that diverted floodwaters into fields. This monumental work was vital to the establishment of China's prosperity. Emperor Yu's efforts, especially his heroic acts in controlling chaos, were praised as essential to establishing cosmic order. What is particularly intriguing is that, while the stories of his irrigation canals have been preserved through time, so too have the stories of how his swaying step—his Yubu—came to be.

Are you still swaying?

Great Yu in Support of Nonnormative Walking/Thinking

What can we learn from the narrative of a mythical figure and his walking step? Embedded within Yu's pursuit of perfection lies a different message conveyed through his swaying step. Emperor Yu performed incredibly demanding work regulating the waters, an endeavor that symbolized taming the chaos of the universe. However, his efforts caused him great physical strain, resulting in an injury that forced him to change the way he walked. This limping, born from the heroic act of flood control, gradually transformed into a ritual.

One could analyze Emperor Yu's step from a strictly technical perspective, focusing on the correct weight shifting, pace, and alignment. Yet what makes his walk so significant is that this limping step, born of necessity, eventually became a ritual movement performed to calm the Big Dipper and maintain the cosmic order. This act, originally a personal adjustment to physical hardship, evolved into a ritual passed down through generations, performed to ensure the stability of the universe (Wang 2018).

The swaying from right to left, aimed at redirecting damage, also seeks to occupy the in-between space—one of ambiguity and shifting perspectives. I call this swaying "esoteric limping" as the stories of Emperor Yu are not only mythical but also deeply mystical. It embodies ancient Chinese ideas about dualities. Yu is depicted as a hydraulic hero with divine power and leadership (Liu, Heping 2012, 93). He performs miracles through his tireless work, while simultaneously laying the foundation for ritualized walking practices. A nondualist interpretation of Emperor Yu's limping—his movement between right and left—suggests a space that resists a fixed, linear path—neither confirming right nor left but instead embracing the in-between.

In a conversation with philosopher Ching-Yuen Cheung (email exchange with author, November 5, 2024), he explained that while the concept of dualities—such as yin/yang, left/right—in Tai Ji originates from the *I Ching* (Book of Changes), Daoism offers an alternative framework. Daoism emphasizes the importance of nothingness, positing that all beings emerge from this formless state. In this sense, Daoism provides an alternative to the binary system of dualities (yin and

Figure 21.2 Please scan the QR code to listen to the instructions. (Skånberg, 2024).

yang), suggesting that the world is not governed solely by opposites but by a deeper unity that transcends them.

When a practice like Emperor Yu's step survives for centuries, remaining vital in practices still connected to its origin, it demonstrates the resilience of oral and embodied knowledge. These forms of knowledge, passed down through generations, persist in their embodied form, countering the misconception that movements or practices cease to exist once the body dies. In this way, the body itself becomes an archive—a living repository of knowledge that can be revisited and reembodied (Baxmann 2007; Lepecki 2010). Through this process, the immaterial nature of certain practices is preserved, allowing them to endure beyond their physical manifestations. "Esoteric limping"—a movement practice that transcends mere physical adjustment and evolves into a cultural and ritualistic act.

Are you swaying or limping?

Esoteric Limping as Coexisting

Please scan the QR code to listen to instructions. You can also read them here:

Take a moment to reflect on bodily pain.
Have you ever walked on crutches?

Do you remember how you had to adjust your walking to the surroundings?
Please close your eyes and remember the experience.
Or stand up and notice the tiny movements that occur
as you pay attention to how your weight shifts.

(Skånberg 2024b)

The story of Emperor Yu also sheds light on how movements, shaped by necessity, come to embody broader social and philosophical meanings. In Yu's case, his limping was not just a physical limitation but became a ritualistic act with societal significance. The narrative of his suffering, caused by the grueling work of flood control, is crucial in justifying his fragility. It elevates his injury, making it part of a heroic and virtuous struggle, which in turn promotes societal acceptance of his condition. This story of hardship and resilience turned the limping step into a powerful symbol—not of weakness but of dedication and sacrifice for the greater good. Is he one of the first historic examples of the workaholic?

But the question arises: How could we accept his limping without these heroic stories? The limping itself is seen not just as a physical necessity but as a form of wisdom—an acceptance of fragility and an embrace of the in between. Growing up in Sweden, I remember how the stars of the Big Dipper were always visible on clear nights. In Swedish folklore, however, the constellation was interpreted as Karlavagnen, or "the Big Wagon." This interpretation of the stars as a wagon led me to imagine Emperor Yu walking with a rollator, gently rocking it back and forth as he moved—calming the sky and, symbolically, calming himself.

The touching part of the Emperor Yu story lies in how his limp became an essential adaptation, necessary for his survival, and continues work. His need to balance and redistribute his weight through limping became an act of resilience and self-care—an act that allowed him to persist in his mission. In this sense, his limp was not a defeat but a process of coexisting with fragility, of finding strength within vulnerability. For any dancer, walker, or person who has tried to continue working while on crutches or enduring pain, Emperor Yu's story resonates deeply, reminding us that our limitations can become a form of wisdom—a place where balance and strength emerge from fragility.

The discussion of limping places Emperor Yu—a mythological figure celebrated for his great organizational skills, and perhaps one of the first historical examples of a workaholic—at the center of this narrative. He was not remembered as someone impaired in his walking but as someone who created his own way of walking out of necessity, and that limping step became a foundational movement for others. I propose the need for a constant curiosity about societal rules, using narrative and embodiment as methods to challenge and transform them. Engaging with how the seemingly "deviating" limping movement feels, creates a deeper understanding and appreciation of its purpose. We should learn to welcome limping, rolling, and swaying as important movement strategies for coexisting. Transforming limping into an essential walking ritual acknowledges mortality, vulnerability, and aging.

Yu's swaying step reflects the consequence of overwork, weariness, and the struggle for balance in a fragile world. The fact that Yubu is still practiced today, despite being nonnormative, invites us to reflect on the possibility of swaying and faltering slowly through society with others. Emperor Yu's step, which evolved into a Daoist walking ritual, demonstrates the importance of choreographed walking for societal engagement. In one of One by Walking's gatherings, I guided participants in contributing creative responses—speculations rather than accurate reenactments—to the walking ritual directed at the Big Dipper (Skånberg 2022a). Each participant was asked to create their own repeatable ten-second walking map within a frame of one square meter. In this way, the act of walking—whether swaying, limping, or moving with intention—becomes not only an individual practice but also a collective one, inviting us to reconsider how we navigate space, engage with one another, and acknowledge the fragility and strength inherent in our shared human experience. Walking, as both ritual and movement, becomes a means of connecting—not only to the world around us but also to each other, in all our imperfect, ever-changing forms.

Please follow these instructions:

Please stand up and try to repeat experiences of arrhythmic walking, limping, or faltering. Imagine that you are influencing the stars as you do it.
How does this thinking affect your movements?
How does this thinking affect your being in the world?

Esoteric Limping as Contemporary Resistance

We rock our newborns in our arms and in cradles to provide comfort and help them fall asleep. Esoteric limping, as a movement practice, is also relevant for our conflicted contemporary times. It could serve as a new strategy for communities to come together.

Since 1997, the term *artivism* has become frequent: "a hybrid neologism that signifies work created by individuals who see an organic relationship between art and activism" (Sandoval, and Latorre 2008, 82–83). The claim is that an artistic experience is transformative and thus changes people and society. In 2005, literature scholar Andrew Hewitt showed through his concept "social choreography" that ideology needs to be understood as something embodied and practiced, not just as an abstract form of consciousness (Hewitt 2005). Choreography is a way of thinking about the relationship of aesthetics to politics. Artivism is something similar to DIY activism. Choreographers emerging from the postmodern era in the United States, such as Deborah Hay, claimed that dance itself can be considered a political activism. However, Hay did not point to waving signs in the streets; her activism was situated in the studio and worked to change mainstream systems of dance training and choreographic transmission (Steinwald 2012). Earlier, choreographer and researcher Pearl Primus (1919–1994) declared her dance to be "the scream which eases for a while the terrible frustration common to all human beings who, because of race, creed or color, are 'invisible'" (Courtney 2021). Later, choreographer Liz Lerman described how it may seem "as if the artist is behaving like an activist, when actually all she is doing is building a world in which she can live and work" (Lerman 2011, 242).

Thus, the final connection I wish to present is between Emperor Yu's esoteric limping and the 1970 Street Action in New York City. Street Action was a protest against the US invasion of Cambodia. Choreographer Yvonne Rainer, a postmodernist icon, was teaching at the School of Visual Arts in New York City. Feeling powerless to continue teaching as if nothing had happened, she gathered her colleagues, Douglas Dunn and Sarah Rudner; friends; and students to walk slowly through the streets of Manhattan's Lower East Side, swaying from side to side with bent necks. Rainer explained that her inspiration came from the movement of factory workers in Fritz Lang's 1927 dystopian film *Metropolis* (Rainer 2013, 340). However, the choreographer of the scenes in *Metropolis* is unknown. I speculate that Fritz Lang (1890–1976), having been injured many times in the First World War, might have embraced movements that did not showcase strength and joy but instead conveyed melancholy and sorrow. The actors portraying factory workers in *Metropolis* offered a silent critique of their time, bowing their heads in deep mourning and slow resistance. In this way, the slow, swaying movements of the factory workers reflected a form of resistance, not through aggressive action but through fragility, vulnerability, and stillness—

much like the esoteric limping of Emperor Yu, which continues to offer a subtle, yet powerful, form of resistance in today's world (Skånberg 2022a).

On January 24, 2024, I recreated a peaceful action indoors with dancers, dance teachers, and choreographers in Stockholm, drawing inspiration from Yvonne Rainer's Street Action as a gesture of hope. I was experiencing the same powerlessness in relation to today's immense global conflicts.

The instructions I gave were:

Stand close to each other, shoulders touching.
Begin to sway together, heads lowered.
When the group agrees on the same rhythm, start moving forward slowly.

After everyone had swayed from side to side and slowly moved forward for a while, I read aloud my written call for peace, interspersed with historical examples of peaceful resistance.

Yvonne Rainer responded to the violence of her time with a choreographed walk, engaging in peaceful protest. At the same time, she created a manifesto of meaning-making through choreography, offering a slow dance that stood in contrast to the more straightforward political marches with signs and placards. I believe this serves as a proposal and a reminder of what we can do through movement. The silent, swaying walking action was able to convey our despair, peace, tears, sadness, and—above all—comfort. Shoulder to shoulder, we created a shared rhythm in silence. Swaying became a meaningful response to a confusing time.

Afterward, we spoke in low voices, sharing thoughts that oscillated between hope and despair. Swaying, as a conversation between the binaries of right and left, found balance in the in between, embodying injustices, disabilities, and brokenness. This exploratory walking style can offer a more inclusive view of physicality and community. Swaying, then, could be both a choreographic method with artistic intent, a means of seeking spiritual guidance, and a movement for consolation with a healing purpose. It was then that I realized that Emperor Yu was not just limping along to mend the world, he was rocking it into peace and contentment.

Just as swaying movements in rituals invite us to embody the in between, the unresolved, and the fragile, Emperor Yu's step reminds us that even in weariness, we can find strength through the act of moving together. Through such practices, we engage with the possibilities of what it means to move together—slowly, thoughtfully, and with intention—toward a more inclusive, balanced, and peaceful world.

References

Ashikaga, Keiko. 1950. "The Bon Festival: A Time of Pleasures and Reflections." *The Journal of Japanese Culture* 5: 228.

Baxmann, Inge. 2007. "The Body as Archive." In *Knowledge in Motion: Perspectives of Artistic and Scientific Research in Dance*, edited by Sabine Gehm, Pirkko Husemann, and Katharina von Wilcke. transcript Verlag.

Boddy, Janice. 1989. *Wombs and Alien Spirits: Women, Men, and the Zar Cult in Northern Sudan*. University of Wisconsin Press.

Courtney, E. 2021. "Pearl Primus on Fighting Ignorance and Prejudice through Dance." *Dance Magazine*.

Dahlstedt, Ami Skånberg. 2022. *Suriashi as experimental pilgrimage in urban and other spaces*, PhD diss. University of Roehampton.

El Guindy, Howaida, and Claire Schmais. 1994. "The Zar: An Ancient Dance of Healing." *American Journal of Dance Therapy* 16 (2): 107–20.

El Hadidi, Hager. 2016. *Zar: Spirit Possession, Music, and Healing Rituals in Egypt*. American University in Cairo Press.

Hewitt, Andrew. 2005. *Social Choreography: Ideology as Performance in Dance and Everyday Movement*. Duke University Press.

Hong, Ge. Ca. fourth century AD. *Baopuzi (Book of the Master Who Embraces Simplicity)* 抱朴子·内篇.

Lancashire, Terence. 2001. "'Kagura'—A 'Shinto' Dance? Or Perhaps Not." *Asian Music* 33 (1): 25–59.

Lepecki, André. 2010. "The Body as Archive: Will to Re-Enact and the Afterlives of Dances." *Dance Research Journal* 42 (2): 28–48.

Lerman, Liz. *Hiking the Horizontal: Field Notes from a Choreographer*. Wesleyan University Press, 2011.

Li, Fuzhong. 2014. "Transforming Traditional Tai Ji Quan Techniques into Integrative Movement Therapy—Tai Ji Quan: Moving for Better Balance." *Journal of Sport and Health Science* 3 (1): 9–15.

Liu, Heping. 2012. "Picturing Yu Controlling the Flood: Technology, Ecology, and Emperorship in Northern Song China." In *Cultures of Knowledge: Technology in Chinese History*, edited by Dagmar Schäfer. Brill.

Ningning, Wang. 2009. *The History of Ancient Chinese Music and Dance* (王宁宁，《中国代乐舞史》). Shanxi People's Publishing House.

Rainer, Yvonne. 2013. *Feelings are facts: a life*. Mit Press.

Sandoval, Chela, and Guisela Latorre. 2008. *Chicana/o Artivism: Judy Baca's Digital Work with Youth of Color*. MacArthur Foundation Digital Media and Learning Initiative.

Skånberg, Ami. 2022a. "Choreography Inspired by Daoist Rituals." Video of One by Walking workshop at Vålådalen Mountain Station, Sweden, September 8. Vimeo, 8:02. https://vimeo.com/818622592.

Skånberg, Ami. 2022b. "Morning Suriashi Walk Led by Ami Skånberg." Video of One by Walking intervention at Vålådalen Mountain Station, Sweden, September 9. Vimeo, 02:10. https://vimeo.com/818625655.

Skånberg, Ami. 2024a Introduction to esoteric swaying. SoundCloud audio. https://soundcloud.com/ami-skanberg/stockholms-konstnaerliga-1

Skånberg, Ami. 2024b. *Reflect on Bodily Pain*. SoundCloud audio. https://soundcloud.com/ami-skanberg/reflectonbodilypain.

Steinwald, Michèle. 2012. "Feminist Movement: Deborah Hay, Artistic Survival, Aesthetic Freedom, and Feminist Organizational Principles." Walker, Minneapolis: Walker Art Center, National Performance Network.

Wang, Xue (王雪). Interview by the author. August 19, 2018.

22

The Pedagogical Praxes of Walking

Simon Poole

Walking as a pedagogical practice, long associated with the Peripatetic school founded by Aristotle in 335 BC, offers a compelling method for integrating physical movement with intellectual and sensory engagement. However, modern education systems largely abandoned walking as a pedagogical tool in favor of sedentary, classroom-based learning models. This shift reflects broader socioeconomic changes during the rise of industrialization and urbanization, when education became formalized to meet the needs of increasingly sedentary, urban populations (Robinson 2010). This transformation is hardly just a Western phenomenon; walking pedagogies are part of a global cultural divide. In many rural or nomadic communities, walking is often central to daily life and learning, facilitating embodied knowledge transmission and cultural continuity (Dyer 2016). In contrast, urbanization and the industrial model of education have prioritized efficiency, progress, and the management of large student populations, reinforcing seated, hierarchical, classroom-based learning as the norm (Dyer 2016; Ingold 2016).

The rise of neoliberal ideologies in the late twentieth century further entrenched these sedentary educational practices, with a heightened focus on measurable outcomes, cost-effectiveness, efficiency, and "impact" (Ball 2003). These values conflict with the open-ended, process-oriented nature of walking, which resists easy quantification (Silver et al. 2020). Neoliberal educational reforms maximize productivity in both time and space, thereby sidelining embodied, mobile learning.

However, critical and decolonial pedagogies have recently revived walking as an "old-new" form of learning, reemphasizing its value in fostering sensory awareness, expanded thinking, and embodied engagement with the environment (Heddon and Myers 2019; McLeod et al. 2020). In recent years, particularly in urban and Western societies, this resurgence can be seen as part of a broader cultural and anthropological reclamation (Lykes et al. 2018; Kondo 2022). Walking has become not just a tool for learning but a sociopolitical act, challenging the alienation caused by urbanization and the hypersedentary nature of modern life. Through walking, learners reclaim their right to engage with their environments, creating new spaces for dialogue, reflection, and resistance (Powell 2019; Langer 1988).

Drawing from various scholarly works, this chapter investigates the concept of walking as a form of pedagogy in its broadest sense, beyond formal schooling environments, emphasizing its role in shaping narratives, temporalities, and embodied experiences (Powell 2019; Blades 2020; Heddon and Myers 2019). It highlights the sensory and material intensities experienced during walking, which transcend mere representation and engage with the environment in a profound manner (Springgay and Truman 2017; Macpherson 2016). The chapter also underscores the significance of walking in fostering a deeper connection with the environment, promoting mindful awareness and expanded listening (Salm and Brogden 2021; Gallagher et al. 2017). Additionally, it explores the cognitive and psychological aspects of walking, such as its meditative and creative dimensions (Keinänen 2015; Matthews et al. 2014).

To develop a categorized structure for considering the various pedagogical praxes, this chapter addresses them as distinct components: the broader context; public pedagogy; pedagogy of affect; learning of environmental connection; convivial pedagogy; and, finally, formal pedagogic structures. This is done purely for ease of interrogation and should be recognized in a sense as somewhat arbitrary. Moreover, many of these categories overlap, run in parallel, or in fact present ambiguous or contradictory positions. Nonetheless, the pedagogical praxes will be contextualized through a breadth of disciplinary engagement, in order to discern any nuanced differences between the approaches. Finally, the historical context and continuity of praxis will then be laid out in the conclusion.

Overall, the chapter provides a comprehensive exploration of walking as a pedagogical practice, elucidating its multifaceted role in intertwining human experiences with the environment. It examines how walking, as a reemerging practice for teaching and learning, intersects with public, environmental, and formal educational contexts. Tracing the evolution of walking pedagogy, the chapter considers its contemporary applications and discusses the challenges and possibilities it presents for both educators and learners in diverse cultural and sociopolitical settings.

The Broader Context

The term "peripatetic" originates from the ancient Greek word *peripatētikós*, meaning "of walking" or "given to walking about" (Oxford Languages 2024). The Peripatetic school, established by Aristotle, was commonly referred to as the Peripatos due to the *peripatoi*, or walkways, of the Lyceum where the school convened (Dyer 2016). Aristotle, unlike Plato, utilized the Lyceum's grounds for gatherings, as he was not an Athenian citizen and could not own property. The school's association with the gymnasium also led to it being known as the Lyceum. The Peripatetic gatherings were likely informal at first, lacking a set curriculum or membership fees and focusing on philosophical and scientific exploration rather than promoting a specific doctrine (Özaykal 2021). While Aristotle himself lectured at the school, research and collaborations among its other members also helped establish its prominence (Bronowski 2019). The school included other notable figures, such as Theophrastus and Eudemus of Rhodes, and its activities were not rigidly structured, allowing for flexibility in research, lectures, and teaching methods involving both junior and senior members. The school's ethos is thought to have emphasized partnership and equality among its members (Smith 2009).

We start with this glance far back in time because walking was a fundamental aspect of the pedagogy of philosophers in the Peripatetic school. Not just a physical exercise, it was a method that allowed for a broader perspective, transcending subjective positions and possibly fostering deeper connections with the environment, in ways that are discussed today (Ingold 2016). While the connection between the Peripatetic school and the natural environment, won by walking, may be a slightly romanticized view, this method certainly promoted a dynamic and flexible approach to learning, emphasizing philosophical and scientific inquiry (Avneri 2022). It also aligned with the standpoints of later philosophers of education, such as John Dewey (1938), and thinkers who continue to expound similar approaches, like Robin Alexander (2020), with his take on dialogic learning. The point is that nowadays walking pedagogy is a method that underscores a broader historical understanding of learning, in which the body actively participates in cognitive processes (Ingold 2016). It is important to note, however, that walking pedagogy is often referred to as "peripatetic pedagogy" and should not be confused with "peripatetic teachers," who are professionals that move from school to school teaching specific subjects, such as music teachers in the context of the United Kingdom.

However, as education became institutionalized during the industrial age, walking as pedagogy diminished, replaced by sedentary classroom learning that prioritized efficiency and measurable outcomes. Indeed, one might even consider the general trajectory of travel over the last couple of hundred years to have been directed by antiwalking pedagogics. This transition reflects a deeper global dichotomy: in many nomadic and rural cultures, learning has historically been a mobile, embodied practice tied to the natural environment (O'Neill et al. 2020). In contrast, urbanization necessitated standardized, sedentary methods that aligned with industrial productivity (Robinson 2010).

The rise of neoliberalism exacerbated this divide, as educational reforms emphasized metrics, accountability, and "impact"—values that marginalized experiential, process-oriented learning methods like walking (Kondo 2022). As a result, sedentary classroom teaching and learning became entrenched, while walking, often perceived as inefficient, was pushed aside in favor of practices that could yield immediate, quantifiable results. However, as the Pink Floyd lyrics "We don't need no education" (Waters 1979) symbolically captured, there has been growing resistance to these rigid educational structures (Keinänen 2015).

Critical pedagogies have begun to challenge these structures, reintroducing walking as a way to disrupt dominant ideologies and engage with learning in embodied, relational ways (Heddon and Myers 2019). In this context, walking serves as a form of public pedagogy, allowing individuals to engage with complex social and environmental issues in ways that transcend traditional classroom boundaries (O'Neill et al. 2023). This chapter situates these contemporary developments within the broader history of walking as a pedagogical tool, drawing connections between past and present practices while acknowledging the sociopolitical forces that continue to shape its role in education.

Walking as Public Pedagogy

As a form of public pedagogy, walking transcends the boundaries of traditional education, engaging individuals with the sociopolitical landscapes through which they move. It enables learners to confront societal structures, power dynamics, and histories embedded within spaces. This form of pedagogy contributes to counter narratives, as individuals embody a critical engagement with their surroundings (Pratt and Johnston 2019; Powell 2019).

For example, Kimberly Powell (2019) discusses walking as a precarious public pedagogy, highlighting the intersection of social and economic structures, spatial politics, and local narratives. This emphasizes how walking can serve as a platform for exploring complex societal issues and historical contexts. Similarly, Nazlı Tümerdem (2019) explores how walking tours in Istanbul, particularly in contested spaces, act as a form of critical public pedagogy by revealing the tensions between development, gentrification, and historical memory. These walking tours disrupt the dominant narratives of the city, fostering a deeper understanding of how urban spaces are shaped by both colonial legacies and neoliberal policies. Both Powell (2019) and Tümerdem (2019) focus on the transformative potential of walking as a critical public pedagogy and show how walking practices can challenge existing power dynamics and hegemonic structures.

In urban contexts, walking tours are often used to critique the capitalist logics that shape cities. As Jonathan Silver et al. (2020) discuss, walking through financialized spaces opens up new avenues for understanding urbanization and its sociopolitical impacts, offering a way to confront the hidden geographies of capitalist development. In this sense, walking becomes a form of resistance against the dominant spatial and temporal structures of the city.

Similarly, Genevieve Blades highlights walking as a wandering ethic that embodies the Freirean notion of a "pedagogy of hope" (2020, 4), suggesting that walking can provide a sense of indeterminacy and vulnerability, leading to new understandings and perspectives. In this sense, walking becomes a form of embodied critique, a way to resist and learn about spatial and ideological confinements.

In the same manner, activist walking tours also illustrate the pedagogical power of walking to engage participants with marginalized histories and identities. For instance, walking has been used to explore queer cultural heritage, offering emotional and embodied experiences that challenge dominant histories (Perryman 2019). In this way, walking can create a relational understanding of place, identity, and memory, encouraging participants to reflect on how public spaces are constructed and who is included in or excluded from these narratives.

However, walking as a public pedagogy is not without its challenges. Walking practices can replicate colonial and exclusionary frameworks, particularly when access to public spaces is unequal or when certain narratives are prioritized over others. As a result, walking must be critically examined in the terms of whose histories are being represented and who has access to participate (Laing 2020). Both Blades (2020) and Tümerdem (2019) caution against romanticizing walking as inherently progressive, urging instead for it to be seen as a tool that demands careful consideration of context and power dynamics.

Despite its potential, walking as public pedagogy also needs to be understood within the limits of its effectiveness. While it provides a unique way to engage with the sociopolitical environment, it is not a remedy for all educational challenges. Walking (as) pedagogy can, under certain conditions, become a part of the very structures it seeks to challenge, and its success relies on critical reflection and the context in which it is employed (Beyes and Steyaert 2020).

In conclusion, walking as public pedagogy offers diverse applications as a method for engaging with complex social issues and fostering critical thinking—from Maggie O'Neill et al.'s (2020) proposition of walking as a critical pedagogy in criminology, which suggests that walking can provide unique insights into crime and justice through corporeal and relational experiences, to activist walking tours that offer a dynamic and embodied approach to learning, engaging participants with the social, cultural, and political aspects of their surroundings. However, as both Blades (2020) and Tümerdem (2019) highlight, walking is not without its limitations. It must be approached critically, with an awareness of its potential to both disrupt and reproduce existing power dynamics.

Walking as a Pedagogy of Affect

Walking as a pedagogy of affect shifts the focus from cognitive processes to emotional and sensory experiences. This perspective suggests that learning is not only a matter of intellectual engagement but also one of feeling, embodied response, and sensory interaction with the world. Scholars have examined walking as an affective practice, particularly in its capacity to provoke emotional intensities and sensory awareness, which contribute to deeper, embodied forms of learning (Beyes and Steyaert 2020; Gallagher et al. 2017).

Affective pedagogies, such as those that incorporate walking, offer learners a way to connect with their surroundings in a more visceral and immediate way. For instance, Timon Beyes and Chris Steyaert (2020) discuss how walking destabilizes conventional knowledge structures by immersing learners in bodily sensations, which leads to a more profound engagement with the learning process. Walking, in this sense, is not merely a means of moving through space but a method of activating affective dimensions often overlooked in traditional pedagogies. Similarly, Emmanouela Mandalaki (2023) explores how walking in fragile environments reveals the vulnerability of both the body and the landscape, making visible the often hidden emotional and sensory connections between learners and their surroundings.

A critical point regarding the assumption that walking always serves as a positive or liberating force in education should be noted here: while walking can indeed foster new emotional and sensory connections, it is not without its limitations. In certain contexts, walking can generate discomfort or unease, particularly when the environments are inhospitable or when learners feel disconnected from the spaces they traverse (Laing 2020). Mandalaki (2023) further highlights the fragility of walking, where the body's physical vulnerability mirrors the uncertainties and risks involved in walking as a pedagogical practice. The affective potential of walking, then, is highly context dependent, requiring careful consideration of how and where it is implemented as a pedagogical tool.

Nonetheless, walking as a pedagogy of affect has been considered within and utilized across a wide range of situations and contexts, spanning all stages of education and both formal and informal settings. Walking's affective power can be seen in formal early childhood education, for example, where walking alongside natural environments such as streams fosters emotional and multisensory engagement in young learners (Lanouette 2022). Such practices not only enhance cognitive development but also help children cultivate emotional connections to the ecological systems around them. Walking, in this context, becomes an affective bridge between the learner and the environment, fostering an embodied understanding that transcends cognitive processing alone. More informally, walking has been employed in museum education to foster emotional connections with artifacts. Mette Muxoll Schou and Anders Sundnes Løvlie propose a shift from the traditional "pedagogy of walking" to a "pedagogy of feeling" (2021, 2) that foregrounds affective engagement with museum collections. This approach highlights how walking can create spaces for learners to emotionally connect with material culture in ways that static viewing cannot (Witcomb 2015).

However, like any pedagogical method, walking presents challenges and risks. While it can foster affective and embodied connections, it can also fail to engage certain learners, especially if the affective dimensions of the practice are not aligned with their personal experiences or if the environments do not support such engagement. Walking should not be regarded as a panacea for all pedagogical challenges. To be effective, it must be critically examined to assess its potential to either enhance or hinder learning, depending on the context (Blades 2020; Beyes and Steyaert 2020).

In conclusion, walking as a pedagogy of affect offers a rich, multisensory approach to learning, engaging both the body and emotions in ways that go beyond cognitive processing. However, its effectiveness is very context dependent, and educators must remain critical of the assumption that walking will always lead to positive affective outcomes (Laing 2020; Mandalaki 2023). By carefully considering the environments and emotional dynamics at play, educators can harness the affective power of walking to cultivate more meaningful and embodied learning experiences.

Walking as the Learning of an Environmental Connection

Incorporating walking into pedagogy potentially offers learners an immersive and embodied way to connect with their environment, fostering sensory and emotional engagement that extends beyond the confines of traditional classroom settings. As a form of environmental pedagogy, walking promotes an experiential approach to learning, encouraging students to engage meaningfully with the physical spaces they inhabit and interact with (Gallagher et al., 2017). Practices such as "listening walks" (2017, 1247) cultivate a heightened sensory awareness, inviting learners to become more attentive to their surroundings and enhancing their understanding of ecological systems.

Through walking, learners also develop a stronger sense of place and belonging, as the act of moving through and interacting with environments can lead to deeper connections with the natural world (Cho and Corkett 2021). This form of learning goes beyond intellectual engagement, incorporating affective responses that play a vital role in fostering environmental stewardship and a commitment to sustainability.

For example, outdoor learning programs that incorporate walking often focus on ecological education, encouraging students to physically engage with their surroundings in order to better understand environmental issues. Holland introduces the concept of a "walking education" (2016, 64), where learners explore natural environments on foot, facilitating a tangible connection with ecological concepts that might otherwise remain abstract in the classroom.

Additionally, Alexandra Fidyk (2021) suggests that walking meditations can cultivate a profound connection with the environment, fostering a deeper appreciation of one's everyday surroundings. Such practices are particularly valuable in urban areas, where access to nature is often limited, making walking an effective means to connect learners with the environment in a direct and meaningful way.

Walking also has the potential to contribute to decolonizing educational practices. By engaging physically with the land, learners are invited to challenge traditional colonial narratives of land ownership and exploitation, nurturing instead a sense of the land as a shared and communal resource. Decolonial walking pedagogies help to shift the focus toward understanding the environment as integral to cultural and ecological sustainability by prioritizing embodied knowledge, storytelling, and engaging with marginalized perspectives. Practices such as group walking, storytelling, and self-disclosure create spaces for learners to reflect on their identities and privilege (Laing 2020). Grounded in praxis that involves continuous movement, reflection, analysis, and action, such pedagogies aim to dismantle injustices by establishing collaborative spaces for knowledge creation and community action (Lykes et al. 2018; Gabi et al. 2022). Learners are encouraged to question traditional narratives, reconsider their histories, and engage in critical resistance. This perspective is vital in addressing the environmental degradation often associated with neoliberal policies, which prioritize resource extraction and economic growth over long-term ecological health. This mode of thinking also intersects with pedagogies of affect. For instance, Silver et al. (2020) argue that walking tours offer a unique opportunity for experiential learning, especially in understanding urbanization processes and capitalist structures.

However, as with other approaches, the effectiveness of walking as an environmental pedagogy depends on the context in which it is implemented. Not all learners may feel comfortable with or connected to natural environments, and for some, walking may not evoke the intended emotional or educational response—echoing the cultural-anthropological dichotomy outlined in the introduction. The success of walking as a pedagogical tool relies on its adaptability to diverse learners and the recognition that environmental connections are deeply personal and context-specific (Sugiyama and Thompson 2007). Therefore while walking can facilitate profound environmental connections, its implementation must consider the individual experiences and accessibility of learners.

Walking as a Convivial Pedagogy

Walking as a convivial pedagogy encompasses various dimensions that intertwine affect, criticality, creativity, and relationality. Scholars have explored walking as a pedagogical practice that extends beyond traditional classroom settings, offering unique opportunities for learning and knowledge creation (Beyes and Steyaert 2020; O'Neill et al. 2023; O'Neill et al. 2020; Mullally et al. 2022). Through walking, individuals engage in affective-pedagogical practices such as straying, drifting, witnessing, and critical reflection, fostering a deeper understanding of the self, others, and the environment (Beyes and Steyaert 2020). This form of critical pedagogy situates cowalkers within complex relations between the body, space, place, senses, and social interactions, emphasizing sustainability, citizenship, and community knowledge (O'Neill et al. 2023). Moreover, walking as a method for teaching, learning, and research enables a corporeal and material understanding of various subjects.

Walking pedagogies also challenge traditional notions of knowledge territorialization, encouraging interdisciplinary conversations and collaborations (Mullally et al. 2022). By engaging in walking practices, individuals reject static forms of knowledge and embrace dynamic, embodied ways of learning that honor diversity and foster connections with the Earth (Salm and Brogden 2021). Furthermore, convivial pedagogic walking facilitates expanded listening, as previously mentioned, offering a unique pedagogical approach to both appreciative and informational listening in educational settings (Gallagher et al. 2017). Pedestrian pedagogy, characterized by improvisatory and relational learning experiences, emphasizes collective movement through space and time, promoting collaborative knowledge construction (Heddon and Myers 2019).

Walking as a convivial pedagogy extends beyond physical movement; it involves (although not automatically) reflective praxis, wonder, and feminist common worldings, activating new possibilities in teacher education and challenging colonial legacies (Franklin-Phipps and Gleason 2019; John 2018). This form of pedagogy encourages educators to embody culturally relevant practices, model diversity, and engage in asset-based teaching approaches (Kondo 2022). Additionally, walking methodologies can contribute to the development of critical and caring practices, fostering attentiveness, responsiveness, and attachments to diverse voices and perspectives (Pratt and Johnston 2019; Wintoneak and Blaise 2021).

In summary, walking as a convivial pedagogy offers a transformative approach to teaching and learning, emphasizing affect, criticality, creativity, and relationality. By engaging in walking practices, individuals can deepen their understanding of themselves, others, and the environment, while challenging traditional educational paradigms and promoting inclusive and sustainable forms of knowledge creation.

Walking in Formal Pedagogic Structures

Walking has traditionally been considered an informal, flexible pedagogical tool, often used in alternative or outdoor learning environments. However, its integration into formal pedagogic structures offers a promising opportunity to enrich traditional education systems. By incorporating walking into more structured educational settings, such as schools and universities, educators can challenge the sedentary nature of classroom learning and foster more embodied, experiential approaches to education (Gallagher et al. 2017).

The integration of walking within formal education models also provides opportunities to rethink how students interact with learning spaces. Pedestrian pedagogy, as discussed by Dee Heddon and Misha Myers (2019), emphasizes learning through collective movement and shared experiences, offering a horizontal approach to knowledge creation that contrasts with the hierarchical nature of traditional classrooms. This approach encourages learners to actively engage in the learning process, promoting collaboration and critical thinking as they move through space and engage with one another.

Walking within formal pedagogic structures, such as teaching lessons on the move or undertaking one-to-one mentoring sessions, can also be aligned with critical pedagogies that challenge traditional models of education. Walking provides an alternative to the rigid, sedentary nature of conventional classrooms, enabling a dynamic and active engagement with educational content. This approach has even been used within the structure of the United Kingdom's Initial Teacher Training framework (Franklin-Phipps and Gleason 2019; Poole 2019). Additionally, walking facilitates an expanded sense of place in learning, where knowledge is not confined to the classroom but instead becomes a part of the students' interactions with their broader environments (Powell 2019).

At the same time, walking in formal education settings presents challenges that must be carefully considered. The structure and organization of formal educational systems, with their focus on standardization, metrics, and outcomes, are at odds with the open-ended and fluid nature of walking as a pedagogical practice (Ball 2003). Integrating walking into these systems requires flexibility and a willingness to embrace more open-ended, process-oriented methods of learning.

For example, walking-based projects in formal university settings have been incorporated into creative curricula, allowing students to explore new forms of knowledge-making (Poole 2024). Luke Jones, Steven Tones, and Gethin Foulkes (2023) describe how walking might foster shared learning in mentor-mentee relationships, and it has also been used in higher education to nurture creativity, reflection, and critical thinking, offering students an opportunity to engage with their learning in more active, interdisciplinary, and participatory ways (Holtham and Owens 2011). Similarly, Gerard Mullally et al. (2022) highlight walking as a tool for reimagining educational spaces, helping to break down the boundaries between classroom and community.

Despite the challenges, the potential of walking to transform formal education structures remains significant. As a pedagogical tool, walking offers a means to disrupt traditional classroom settings, promote collaborative learning, and encourage a more holistic understanding of educational content.

Conclusions

The integration of walking into pedagogical practices reveals its potential to serve as a powerful tool for learning, offering embodied, affective, and sensory experiences that complement traditional cognitive approaches. Throughout this chapter, the various dimensions of walking pedagogy have been explored, demonstrating how walking serves at once as a form of public pedagogy, a method of environmental connection, and a critique of sedentary educational models. Each of these perspectives contributes to a more holistic understanding of education—one that recognizes the importance of movement, space, and the body in the learning process (Kraftl et al. 2022; Gabi et al. 2022).

The resurgence of walking in contemporary pedagogical practices—particularly in urban, Western, and Global North contexts—can be understood not only as a response to the constraints of modern education but also as part of a broader cultural, anthropological, and folkloric phenomenon. In these contexts, walking can be seen as a means of reclaiming space and discourse, a response to the increasing alienation and fragmentation caused by urbanization and neoliberal policies. Walking, in this sense, is not just a pedagogical tool but also a sociopolitical act—a way to reconnect with community, environment, and self in a world that prioritizes speed, efficiency, and digital interaction (Powell 2019). Whether consciously or unconsciously, the act of walking revives traditions of embodied knowledge, offering a form of resistance to the hypersedentary lifestyles imposed by modern urban life.

As a result, walking as a public pedagogy allows learners to engage with sociopolitical issues, fostering critical reflection and the potential for counter-narratives (Powell 2019). In environmental education, walking facilitates a direct and emotional connection to the natural world, helping learners develop a deeper sense of place and sustainability (Laing 2020). Within formal pedagogic structures, walking provides a means of disrupting traditional hierarchies and promoting collaborative, experiential learning (Heddon and Myers 2019).

However, as discussed, incorporating walking into education is not without its challenges. Its effectiveness as a pedagogical tool is highly context dependent, requiring careful consideration of learners' needs, environments, and access to space (Sugiyama and Thompson 2007; Bergen et al. 2022). Additionally, while walking can promote critical reflection and engagement, it can also reproduce exclusionary practices if not implemented thoughtfully. For instance, access to safe and walkable spaces is not universally available, and certain populations may find walking difficult or uncomfortable due to physical, social, or economic barriers (Laing 2020).

Despite these challenges, and indeed, perhaps because of the transformative thinking that such challenges inspire or require, walking remains a valuable pedagogical tool, offering the potential to transform how we think about learning, space, and the body. As educators continue to explore the possibilities of walking in their pedagogical practices, the emphasis must remain on inclusivity, adaptability, and critical engagement with the sociopolitical and environmental contexts in which learning takes place. Walking offers a pathway to more holistic and engaged forms of education, where knowledge is created through movement, interaction, and lived experience.

References

Alexander, Robin. 2020. *A Dialogic Teaching Companion*. Routledge.

Avneri, Ira. 2022. "At the Feet of Philosophy: The Dialectics of the Two-Legged Thinker." *Philosophy and Literature* 46 (2): 312–33.

Ball, Stephen J. 2003. "The Teacher's Soul and the Terrors of Performativity." *Journal of Education Policy* 18 (2): 215–28. 10.1080/0268093022000043065.
Bergen, Kyle, Megan Jubenvill, Kyle Shaw, Erin Steen, Hal Loewen, Sophia Mbabaali, and Ruth Barclay. 2022. "Factors Associated with Outdoor Winter Walking in Older Adults: A Scoping Review." *Canadian Journal on Aging / La Revue Canadienne Du Vieillissement* 42 (2): 316–27. https://doi.org/10.1017/s0714980822000460.
Beyes, Timon, and Chris Steyaert. 2020. "Unsettling Bodies of Knowledge: Walking as a Pedagogy of Affect." *Management Learning* 52 (2): 224–42. https://doi.org/10.1177/1350507620979713.
Blades, Genevieve. 2020. "Walking as a Wandering Ethic of (Re)Location: A Public 'Pedagogy of Hope.'" *Journal of Public Pedagogies* (5): 4–19. https://doi.org/10.15209/jpp.1210.
Bronowski, Ada. 2019. *The Stoics on Lekta: All There Is to Say*. Oxford University Press. https://doi.org/10.1093/oso/9780198842880.003.0007.
Cho, Christine, and Julie Corkett. 2021. "Gateways to New Learnings: Walking to Re/Vision an International Practicum." *Journal of the Canadian Association for Curriculum Studies* 18 (2): 150–62. https://doi.org/10.25071/1916-4467.40587.
Dewey, John. 1938. *Experience and Education*. Collier-MacMillan.
Dyer, Caroline. 2016. "Evolving Approaches to Educating Children from Nomadic Communities." *Prospects* 46 (1): 39–54. https://doi.org/10.1007/s11125-016-9381-6.
Fidyk, Alexandra. 2021. "Walking Meditations: Becoming Place, Place Becoming." *Journal of the Canadian Association for Curriculum Studies* 18 (2): 103–18. https://doi.org/10.25071/1916-4467.40660.
Franklin-Phipps, Asilia, and Tristan Gleason. 2019. "Walking Methodologies with/in Teacher Education." *Journal of Public Pedagogies* (4). https://doi.org/10.15209/jpp.1193.
Gabi, Josephine, Anna Olsson Rost, Diane Warner, and Uzma Asif. 2022. "Decolonial Praxis: Teacher Educators' Perspectives on Tensions, Barriers, and Possibilities of Anti-Racist Practice-Based Initial Teacher Education in England." *The Curriculum Journal* 34 (1): 83–99. https://doi.org/10.1002/curj.174.
Gallagher, Michael, Jonathan Prior, Martin Needham, and Rachel Holmes. 2017. "Listening Differently: A Pedagogy for Expanded Listening." *British Educational Research Journal* 43 (6): 1246–65. https://doi.org/10.1002/berj.3306.
Heddon, Dee, and Misha Myers. 2019. "Pedestrian Pedagogy: The Walking Library for Women Walking." *Journal of Public Pedagogies* (4). https://doi.org/10.15209/jpp.1179.
Holland, LeAnn. 2016. "Reconsidering the 'Ped' in Pedagogy: A Walking Education." *Philosophy of Education* 72: 64–73. https://doi.org/10.47925/2016.064.
Holtham, Clive, and Allan Owens. 2011. *Using the Urban to Span the Boundaries Between Diverse Disciplines: Drama Education and Business Management, Practice*

and Evidence of Scholarship of Teaching and Learning in Higher Education. Vol. 6, (3) ECE Conference Special Issue December, pp. 292–305.

Ingold, Tim. 2016. "On Human Correspondence." *Journal of the Royal Anthropological Institute* 23 (1): 9–27. https://doi.org/10.1111/1467-9655.12541.

John, Rebecca. 2018. "Unsettling the Colonial Places and Spaces of Early Childhood Education." *Contemporary Issues in Early Childhood* 20 (2): 213–15. https://doi.org/10.1177/1463949118771499.

Jones, Luke, Steven Tones, and Gethin Foulkes. 2023. "Talking the Talk: Dialogic Mentoring in Physical Education." *Journal of Physical Education, Recreation & Dance* 94 (5): 40–45. https://doi.org/10.1080/07303084.2023.2185328.

Keinänen, Mia. 2015. "Taking Your Mind for a Walk: A Qualitative Investigation of Walking and Thinking among Nine Norwegian Academics." *Higher Education* 71 (4): 593–605. https://doi.org/10.1007/s10734-015-9926-2.

Kondo, Chelda Smith. 2022. "Walking the Talk: Employing Culturally Relevant Pedagogy in Teacher Education." *Teachers College Record* 124 (4): 65–94. https://doi.org/10.1177/01614681221096797.

Kraftl, Peter, Marcia McKenzie, Kalervo Gulson, et al. 2022. "Learning Spaces: Built, Natural and Digital Considerations for Learning and Learners." In *Reimagining Education: The International Science and Evidence based Assessment*, edited by Anantha K. Duraiappah, Nienke M. van Atteveldt, Joanne M. Buil, Kriti Singh, and Rongxiu Wu. UNESCO MGIEP. https://doi.org/10.56383/kwaw2002.

Laing, Anna. 2020. "Decolonising Pedagogies in Undergraduate Geography: Student Perspectives on a Decolonial Movements Module." *Journal of Geography in Higher Education* 45 (1): 1–19. https://doi.org/10.1080/03098265.2020.1815180.

Langer, Beryl. 1988. "The Practice of Everyday Life." *Contemporary Sociology a Journal of Reviews* 17 (1): 122–24. https://doi.org/10.2307/2069486.

Lanouette, Kathryn. 2022. "Emotion, Place, and Practice: Exploring the Interplay in Children's Engagement in Ecologists' Sampling Practices." *Science Education* 106 (3): 610–44. https://doi.org/10.1002/sce.21702.

Lykes, M. Brinton, Courtnye R. Lloyd, and Karla M. Nicholson. 2018. "Participatory and Action Research Within and Beyond the Academy: Contesting Racism Through Decolonial Praxis and Teaching 'Against the Grain.'" *American Journal of Community Psychology* 62 (3–4): 406–18. https://doi.org/10.1002/ajcp.12290.

Macpherson, Hannah. 2016. "Walking Methods in Landscape Research: Moving Bodies, Spaces of Disclosure and Rapport." *Landscape Research* 41 (4): 425–32. https://doi.org/10.1080/01426397.2016.1156065.

Mandalaki, Emmanouela. 2023. "Breaking the Plaster: Making Do with the Fragility of the Body." *Gender Work and Organization* 30 (4): 1450–67. https://doi.org/10.1111/gwao.12996.

Matthews, Douglas R., Valerie A. Ubbes, and Valeria J. Freysinger. 2014. "A Qualitative Investigation of Early Childhood Teachers' Experiences of Rhythm as Pedagogy."

Journal of Early Childhood Research 14 (1): 3–17. https://doi.org/10.1177/1476718x14523745.

McLeod, Kim, Robyn Moore, Duncan Robinson, Derya Ozkul, Sarah Ciftci, and Kate Vincent. 2020. "Using the Pluriverse Concept to Critique Eurocentrism in Education." *Journal of Applied Learning & Teaching* 3 (Special Issue). https://doi.org/10.37074/jalt.2020.3.s1.4.

Mullally, Gerard, Maggie O'Neill, Deirdre de Bhailís, Brendan Tuohy, Maggie Breen, Andrew Duggan, and Elaine Ní Loinsigh. 2022. "Walking, Talking, [Re-]Imagining Socio-Ecological Sustainability: Research on the Move/Moving Research." *Irish Journal of Sociology* 31 (1): 37–62. https://doi.org/10.1177/07916035221118023.

Nash, Louise. 2018. "Performing Place: A Rhythmanalysis of the City of London." *Organization Studies* 41 (3): 301–21. https://doi.org/10.1177/0170840618789161.

O'Neill, James, and Maggie O'Neill. 2021. "Walking in the Boboli Gardens in Florence: Toward a Transdisciplinary, Visual, Cultural, and Constellational Analyses of Medieval Sensibilities in the *Hypnerotomachia Poliphili*." *Qualitative Inquiry* 28 (2): 267–80. https://doi.org/10.1177/10778004211042357.

O'Neill, Maggie, Claire Edwards, Caitríona Ní Laoire, Ger Mullally, and Mastoureh Fathi. 2023. "Introduction to the Special Issue." *Irish Journal of Sociology* 31 (1): 3–11. https://doi.org/10.1177/07916035221126709.

O'Neill, Maggie, Ruth Penfold-Mounce, David Honeywell, Matt Coward-Gibbs, Harriet Crowder, and Ivan Hill. 2020. "Creative Methodologies for a Mobile Criminology: Walking as Critical Pedagogy." *Sociological Research Online* 26 (2): 247–68. https://doi.org/10.1177/1360780420922250.

Oxford Languages. 2024. Retrieved from https://www.google.com/search?q=peripatetic+meaning&oq=peripatetic&gs_lcrp=EgZjaHJvbWUqDQgBEAAYgwEYsQMYgAQyDwgAEEUYORiDARixAxiABDINCAEQABiDARixAxiABDIHCAIQABiABDIHCAMQABiABDIHCAQQABiABDIHCAUQABiABDIHCAYQABiABDIHCAcQABiABDIHCAgQABiABDIHCAkQABiABNIBBzk4MWowajeoAgCwAgA&sourceid=chrome&ie=UTF-8.

Özaykal, Kayhan Ali. 2021. "Philosophical Theology in Islam: Later Asharism East and West." *Ilahiyat Studies* 12 (1): 126–31. https://doi.org/10.12730/13091719.2021.121.222.

Perryman, Georgina. 2019. "Queering Pride: Walking Towards a Queer Future in Ireland." *Journal of Public Pedagogies* (4). https://doi.org/10.15209/jpp.1180.

Poole, Simon. 2019. *The Walk to Kitty's Stone*. Soil Records.

Poole, Simon. 2024 [2017]. *Playing with Ekphrasis*. WeaselBear Publishing.

Powell, Kimberly. 2019. "Walking as Precarious Public Pedagogy." *Journal of Public Pedagogies* (4). https://doi.org/10.15209/jpp.1188.

Pratt, Susanne, and Kate Johnston. 2019. "Speculative Harbouring: Wading into Critical Pedagogy and Practices of Care." *Journal of Public Pedagogies* (4). https://doi.org/10.15209/jpp.1183.

Robinson, Ken. 2010. "Changing Education Paradigms." RSA Animate. YouTube. https://www.youtube.com/watch?v=zDZFcDGpL4U&t=16s.
Salm, Twyla, and Lace Marie Brogden. 2021. "The Noise of Walking." *Journal of the Canadian Association for Curriculum Studies* 18 (2): 64–79. https://doi.org/10.25071/1916-4467.40613.
Schou, Mette Muxoll, and Anders Sundnes Løvlie. 2021. "The Diary of Niels: Affective Engagement Through Tangible Interaction with Museum Artifacts." In *Digital Heritage. Progress in Cultural Heritage: Documentation, Preservation, and Protection. EuroMed 2020. Lecture Notes in Computer Science*, Vol. 12642, edited by Marinos Ioannides, Eleanor Fink, Lorenzo Cantoni, and Erik Champion. Springer. https://doi.org/10.1007/978-3-030-73043-7_24https://doi.org/10.48550/arxiv.2011.11375.
Silver, Jonathan, Desiree Fields, Rich Goulding, Isaac Röse, and Siobhan Donnachie. 2020. "Walking the Financialized City: Confronting Capitalist Urbanization through Mobile Popular Education." *Community Development Journal* 56 (1): 161–79. https://doi.org/10.1093/cdj/bsaa044.
Smith, Richard. 2009. "Between the Lines: Philosophy, Text and Conversation." *Journal of Philosophy of Education* 43 (3): 437–49. https://doi.org/10.1111/j.1467-9752.2009.00700.x.
Springgay, Stephanie, and Sarah Truman. 2017. "A Transmaterial Approach to Walking Methodologies: Embodiment, Affect, and a Sonic Art Performance." *Body & Society* 23 (4): 27–58. https://doi.org/10.1177/1357034x17732626.
Springgay, Stephanie, and Sarah Truman. 2019. "Walking in/as Publics: Editors Introduction." *Journal of Public Pedagogies* (4). https://doi.org/10.15209/jpp.1170.
Sugiyama, Takemi, and Catharine Thompson. 2007. "Outdoor Environments, Activity and the Well-Being of Older People: Conceptualising Environmental Support." *Environment and Planning A: Economy and Space* 39 (8): 1943–60. https://doi.org/10.1068/a38226.
Tümerdem, Nazlı. 2019. "The School of Site: Istanbul Walkabouts." *Journal of Public Pedagogies* (4). https://doi.org/10.15209/jpp.1190.
Waters, Roger. 1979. *Another Brick in the Wall. On The Wall.* Columbia Records.
Wintoneak, Vanessa, and Mindy Blaise. 2021. "Voicing Derbarl Yerrigan as a Feminist Anti-Colonial Methodology." *River Research and Applications* 38 (3): 435–42. https://doi.org/10.1002/rra.3822.
Witcomb, Andrea. 2015. "Toward a Pedagogy of Feeling: Understanding How Museums Create a Space for Cross-Cultural Encounters." In *Museum Theory. Part 2: Disciplines and Politics*, edited by Andrea Witcomb and Kylie Message. John Wiley & Sons. https://doi.org/10.1002/9781118829059.wbihms116.

Index

Contributors

Camilla Brudin Borg (University of Gothenburg, Sweden) is a literary scholar (Docent) and one of the initiators and conveners of the One by Walking research network. Specializing in literature, ecocriticism, and environmental humanities, she is currently exploring walking narratives along the Camino de Santiago de Compostela in Spain and other long-distance hiking trails, combining ecocritical literary analysis with walking and spatiality methods.

Anna Viola Hallberg is a Stockholm-based artist and curator with an MSc in international museum studies. She is the curator of the art node BKN and cocurator of Walking as Practice. She creates lens-based and embodied art, focusing on rural art concepts and social justice.

Emily Höckert (University of Lapland, Finland) is a tourism researcher and member of the Intra-Living in the Anthropocene (ILA) research group. Their work cultivates relational ways of being, walking, and storying with multispecies communities in everyday, proximate settings. Höckert is currently working in the Profi7 Multispecies Hospitality project, funded by the Research Council of Finland.

Jannicke Høyem (Norwegian School of Sport Sciences) is professor of *friluftsliv* (outdoor education), with research interests in didactics for adults in outdoor settings, practical knowledge, health and well-being, sustainability, and human-nature relationships. Høyem has extensive experience working with journeying in the winter mountains and exploring how these experiences can be conveyed and understood.

Rose Keller (Norwegian Institute for Nature Research) is a research scientist specialized in nature-based tourism in protected areas. With a background in acoustic assessments of national parks, as well as visitor management and conservation governance at the National Park Service (Alaska), she currently leads a national project mapping natural soundscapes and noise in Norwegian national parks, along with the Nordic Soundscapes Research Network.

Veera Kinnunen (University of Lapland, Finland) is an expert on more-than-human relations with waste, soil, and microbes. A lecturer in sociology and Docent of environmental social sciences and humanities, she is currently affiliated with the University of Oulu's transdisciplinary research program Biodiverse Anthropocenes (ANTS) and is the co-PI of More Than Just Human Homes (HOMINGS, RCF 2024–2028).

Alice Lewis (RMIT University, Australia) is a landscape architecture academic investigating how the power of prosthetic technologies can be redirected or reimagined to reveal often hidden capitalist legacies of landscape destruction and to design more regenerative modes of inhabitation. She is the program director of the Master of Landscape Architecture at RMIT University.

Karen V. Lykke (University of Oslo, Norway) is an agronomist and ethnologist with a PhD in cultural history and is a professor at Oslo's Centre for Development and the Environment. Her research explores landscape studies and the social and cultural aspects of food and fodder systems, with ongoing research projects on the practice and science of regenerative farming.

John Martin (University of Plymouth, United Kingdom) is an interdisciplinary academic with a background in geography, specializing in the use and development of collaborative tools and holistic approaches to address real-world global challenges.

Roger Norum (University of Oulu, Finland) is a social anthropologist and occasional walker who researches mobility, media, and the environment across the Arctic and South/Southeast Asia. He currently directs two major research projects: FORbEST, which focuses on forest management practices, and RURALEX, which examines the crisis of environmental expertise in rural Europe.

Ane Ohrvik (University of Oslo) has a background in folkloristics and is a professor of cultural history, specializing in the history of knowledge. She studies beliefs, rituals, narratives, and knowledge production in areas ranging from the witch hunts of early modern Europe to contemporary pilgrimage in the Nordic countries.

Maria Papadomanolaki (Greece) is a sound artist and researcher whose work examines the role of sound and walking in shaping how we perceive, create memories, and make place and identity within our environments. She is the cofounder of the Sound Camps arts cooperative and a member of the Shortwave Collective.

Tina Paphitis (University of Bergen, Norway) is a folklorist with a background in archaeology and critical heritage studies, working in the field of environmental humanities. Her research interests include legends and landscapes of Britain and the Nordic region, environmental folklore, walking methodologies and experiential fieldwork, as well as folklore and archaeology in fantasy literature.

Kayla Parker (University of Plymouth, United Kingdom) is an artist filmmaker whose research centers on place, landscape, and memory. Working with material film and digital technologies, she employs a process-based, dialogic methodology informed by *écriture féminine*, investigating the interrelationship between bodies and forgotten, liminal spaces, framed by subjectivity, embodiment, and technological mediation.

Jesse D. Peterson (University College Cork, Ireland) is a lecturer with the Radical Humanities Laboratory and Department of Geography. He researches relationships among environmental change, culture, and science and technology using a postdisciplinary approach. He has contributed to fields such as geography, environmental humanities, death studies, history, and the natural sciences, working on topics including ocean pollution and degradation, ethics in mariculture, biodiversity infrastructures and data, citizen science, algae, and creative research methods.

Simon Poole (University of Chester, United Kingdom) is an associate professor of cultural education and codirector of the Centre for Research into Education and Creativity through Arts Practices (RECAP). As a folklorist—and the holder of the National Collection of Mentha in the United Kingdom—his work focuses on creative practices, including singing, songwriting, storytelling, walking, poetry, and, most recently, gardening.

Outi Rantala (University of Lapland, Finland) is a tourism researcher from the Intra-Living in the Anthropocene (ILA) research group. Together with her colleagues, she explores walking and skiing methods to foster caring engagement with proximate surroundings, beings, and thoughts. Rantala is currently leading the Profi7 Multispecies Hospitality project, funded by the Research Council of Finland.

Alejandro Reig (University of Oulu, Finland, and University of Oxford, United Kingdom) is a social anthropologist researching socioenvironmental issues in Amazonia and the Mediterranean, focusing on landscape, place-making, and various forms and scales of mobility. He combines his research work with consultancy, writing, and teaching anthropological theory and methods.

Noel B. Salazar (KU Leuven, Belgium) is a social and cultural anthropologist studying polycrises, planetary health, endurance, wisdom, imaginaries, heritage, and travel. He is the founder of the EASA Anthropology and Mobility Network (AnthroMob) and the Cultural Mobilities Research (CuMoRe) cluster at KU Leuven.

Dani Schrire (Hebrew University of Jerusalem, Israel) is a folklorist and scholar of everyday culture, folkloristics, and the practices of collecting and archiving. When he is not walking in the streets of Jerusalem, he is a co-PI in a VW-funded project on postcards and postcarding in times of upheaval, as well as a co-PI in the International Research Training Group on Belongings: Jewish Material Culture in Twentieth-Century Europe and Beyond.

Oula Seitsonen, Sakarin-Pentin Ilarin Oula (University of Oulu, Finland) is an archaeologist and geographer with a broad interest in all things, humans, and landscapes, both past and present. He is the PI of two projects: "Contemporary Archaeological Perspectives on the Inequality of the Welfare Society" (Kone Foundation, 2022–2027) and "Eerie Arctic Aviation: Impacts and Cultural Legacies of 20th-Century Military Aviation in the Circumpolar Zone" (Research Council of Finland, 2024–2028).

Sanna Seitsonen is a zooarchaeologist with Arkteekki Ltd. She has carried out archaeological and anthropological fieldwork in Finland, Norway, Russia and Canada. Sohvi Seitsonen, Elsa Seitsonen and Elvi Seitsonen are the daughters of Sanna and Oula and co-writers to chapter 15.

Joane Serrano (University of the Philippines Open University) is chancellor of the university and professor of environmental and sustainability communication. She has a wide array of research interests, including sustainability, environmental communication, Indigenous knowledge, experiential learning, and open and distance eLearning (ODeL). She is also currently the editor-in-chief of the UPOU-managed *Journal of Management and Development Studies.*

Ami Skånberg (Stockholm University of the Arts and University of Gothenburg, Sweden) is a performer and artistic researcher with a PhD in dance from the University of Roehampton. Her research interests include practice-led and concerned walking, Japanese dance, screendance, gender-codified movement practices, and auto-ethnographic accounts from within the practice. She is the vice chair of the Nordic Forum for Dance Research.

Daniel Svensson (Malmö University, Sweden) is an associate professor in sport science. He has a keen interest in issues relating to knowledge, technology, and performance in sport and outdoor recreation and has conducted research on the balance between scientific and practical knowledge for coaches and athletes. Svensson is also a Swedish Olympic Committee Research Fellow.

Daniele Valisena (GHI Washington Pacific Office, University of California at Berkeley, United States) is an environmental historian working at the intersection of migration studies, environmental humanities, and history of science and technology. His research incorporates walking ethnographies and geohistory, tracing the trajectories of Italian mobilities, fascism, and cattle genomics.

Dominica Williamson is an artist and designer specialising in arts-led research in the landscape. Her work, deeply rooted in sustainability, spans client and self-initiated, transdisciplinary projects that engage with embodiment through walking and making. She promotes sensual, more-than-human engagement through methods she's co-created, such as the 'Mini-Landscape'. Clients include Plymouth University, Exeter University, Plymouth Marine Laboratory, and various NGOs.

Hanna Elisabet Åberg (Uppsala University, Sweden) is an avid walker, landscape architect, and planner. Having worked across the disciplines of human geography, environmental psychology, and urban planning, her research engages with the intersections of rural landscape, critical heritage, place identity, and environmental social science.

Susanne Österlund-Pötzsch is a researcher (Docent) in folkloristics at Åbo Akademi University and an archivist specializing in tradition materials at the Society of Swedish Literature in Finland. Walking, as a practice, experience, and form of expression, has long been an area of research and fascination for her.

www.ingramcontent.com/pod-product-compliance
Ingram Content Group UK Ltd.
Pitfield, Milton Keynes, MK11 3LW, UK
UKHW020221290926
15470UKWH00001B/315

9 798216 365624